Plants and Man on the Seychelles Coast

A STUDY IN HISTORICAL BIOGEOGRAPHY

Plants and Man on the Seychelles Coast

A STUDY IN HISTORICAL BIOGEOGRAPHY

Jonathan D. Sauer

MADISON, MILWAUKEE, AND LONDON

THE UNIVERSITY OF WISCONSIN PRESS · 1967

Published by
The University of Wisconsin Press
Madison, Milwaukee,
and London
U.S.A.:
Box 1379,
Madison, Wisconsin 53701
U.K.:
26–28 Hallam St.,
London, W. 1
Copyright © 1967 by the Regents
of the University of Wisconsin
Printed in
the United States of America
by the North Central Publishing Co.,
St. Paul, Minnesota
Library of Congress Catalog
Card Number 67-13556

To a real salty beachcomber — my mother

Preface

Anyone interested in the natural history of islands is bound sooner or later to come upon the special problem of the Seychelles. The archipelago lies in the Indian Ocean, over six hundred miles north-northeast of Madagascar, but is continental in its geology, being formed of granites unique among oceanic islands. This anomaly was clearly stated in Alfred Russel Wallace's classic book, *Island Life*, and has played a crucial role in theories of Indian Ocean land bridges and shifting continents. The native biota, a puzzling mixture of endemic and cosmopolitan species, has also fed much speculation about ancient migrational pathways.

My approach to the Seychelles was long and rambling. An old concern with origins of some tropical crops and volunteer weeds of artificial habitats led me to study their wild relatives growing as pioneers along Caribbean and Gulf of Mexico beaches. I was soon inescapably and happily involved with tropical coastal vegetation in general, finding a world of problems in its geographic patterns. Along almost any tropical shoreline, sharp microclimatic gradients combine with edaphic contrasts between tidal swamps, sea cliffs, dunes, and beach ridges to make intricate local borders. These are often dynamic, readjusting endlessly to seasonal and nonperiodic changes in wind, spray, and wave reach. Superimposed on everything else are broad climatic transitions and historical differences between regional floras. Certain coastal species poorly adapted for long-range dispersal were originally confined to limited regions; with increasing human dispersal, many of these have become widely naturalized and some are still actively colonizing new territory. Other species were effectively dispersed by ocean currents from island to island and continent to continent; some of these circumnavigated the globe and filled their potential areas long ago.

In 1958, when I was just starting to follow these plants, my West Indian itinerary happened to overlap that of a Coastal Studies Institute team led by Richard Russell. Out of this lucky meeting came the privilege of further field work in various parts of the world with Russell, William McIntire, and other geographers of the C.S.I. staff, including three seasons in the Indian Ocean region. Before reaching the Seychelles, we had studied the ocean's continental coasts in Australia and Africa and some of its islands, particularly volcanic Mauritius and Darwin's classic atoll, Cocos-Keeling.

Finally, in late April of 1963, McIntire and I arrived in the Seychelles by the monthly passenger ship from Mombasa. Passages are hard to get and we could originally arrange only a one-month's stay, which we extended a few weeks by returning on an unscheduled freighter in mid-June. This short time was enough for my simple original purposes: to see what role the native endemics played on the coast and how seashore species more familiar to me behaved in a strange setting. Only one endemic proved important; the granite substrate, high-tidal range, and local climatic regime showed some nice effects on plant microdistributions, but they were fairly easy to characterize by transecting and mapping sample areas. Interactions between sea, air, and land give plenty to think about on any coast and the exceptional beauty of the Seychelles made poking along the shoreline a constant delight, but the vegetation looked disappointing at first. Only scattered natural stands have been left intact and extensive areas are blanketed by what seemed to be ordinary coconut plantations with a monotonous weedy undergrowth.

It eventually dawned on me that this apparently uninteresting vegetation had a history that was extraordinary both in itself and in its documentation. In the first place, these extremely ancient islands were among the last to be discovered and settled by man. Thus the whole process of transforming a completely wild to a largely tame landscape was compressed into two centuries, during which much was recorded by a succession of remarkable observers. Also, plant introductions to the colony came through narrow and simply traced channels. Most unexpected of all, the coconuts turned out to be not historic introductions but peculiar native varieties independently domesticated in the archipelago.

I began piecing together vegetational history from records available in the Seychelles and Mauritius, later filling in from published material in the British Museum and several U.S. libraries. In the body of this book, historic changes are presented chronologically, after a deductive reconstruction of primeval vegetation and before a first-hand account of present vegetation. The focus throughout is coastal, the discrete story of interior vegetation being peripherally noted. The interior was originally dominated by curious endemics that suffered drastically under human invasion and have long since been reduced to vanishing relics. Previous botanical work on the Seychelles has tended to concentrate on the rare inland natives, and little can now be added to their sad story.

The cast of botanical characters includes 119 coastal species, each supposedly a breeding population and a biological unit in evolution and migration. Of these, 57 are considered to have been present aboriginally and 62 to have been introduced artificially. In the text, species are called by their common names if they are widely known, otherwise by their generic names. Appendices contain the lists of scientific and vernacular names, together with keys to symbols used in the text.

Voucher collections are deposited in the University of Wisconsin Herbarium and duplicates of some are in the herbaria of the University of California, the Field Museum of Natural History, and the Missouri Botanical Garden.

J.D.S.

Madison, Wisconsin
November 1966

Acknowledgments

The field investigation was sponsored by the Coastal Studies Institute, Louisiana State University, with financial support from the Geography Branch, Office of Naval Research. McIntire worked mainly on geomorphology, the subject of a separate report that Russell and McIntire published in 1965; but he also contributed observations and ideas on the vegetation and was largely responsible for our logistics and official arrangements. His Excellency the Governor, the Earl of Oxford and Asquith, took a personal interest in our studies and sanctioned generous help by government staff. J. F. G. Lionnet, Director of Agriculture, provided laboratory facilities, access to agricultural records, and expert counsel, as did members of his staff: S. M. Savy, Agricultural Officer; R. Dubuisson, Technical Officer; and C. J. Piggott, Agronomist, who also conducted me on a pirogue venture to Ile aux Vaches Marines. A. W. T. Webb, Government Archivist, patiently instructed me in some elements of Seychelles history and brought forth invaluable documentary records. A. Penrose, Director of

Medical Services, arranged for vehicle transport and lodging on Praslin and boat transport from there to Curieuse and La Digue. Temooljee & Co., Ltd., H. Savy, and T. d'Offay volunteered boat transportation and hospitality on their islands, Silhouette, Frigate, and Sud-est. On Mahé, Mrs. P. S. Le Geyt, R. Deltel, M. G. Boullé, and others tolerated my invasions of their coastal properties with great good nature. Colonel E. E. Lasch, U.S.A.F., arranged our passage back to Mombasa on a ship chartered to the U.S. Tracking Station on Mahé.

Library work in the British Museum was supported by the Geography Branch of the Office of Naval Research and by the Whitbeck Travel Fund administered by the Department of Geography, University of Wisconsin. Maps and profile diagrams were prepared in the University of Wisconsin Cartographic Laboratory by Jeanne Tsou Liu and D. Woodward, respectively, under the direction of R. Sale.

Contents

xiv

CONTENTS

Illustrations

Figures

Plants and Man on the Seychelles Coast

A STUDY IN HISTORICAL BIOGEOGRAPHY

Natural 1
Setting

The Seychelles proper* include about two dozen islands and islets and a few isolated rocks, all emerging from a great shallow bank. Bird and Denis islands are low piles of coral debris near the northern edge of the bank, about thirty miles beyond the northern limit of the map (Fig. 1). All the rest, grouped compactly near the center of the bank, are granitic, formed of jointed intrusive rocks that crystallized at great depth and were later cut by vertical dikes of basalt. Recent radiometric determinations at the University of Cambridge date the granites which form Mahé, Praslin, and some other islands as pre-Cambrian, the syenites and other granitic rocks of Silhouette as early Tertiary (Miller and Mudie, 1961; Baker, 1963; Browne, 1963; Matthews, 1965).

---

* Not included in the present study are the Amirantes, Aldabra, and other widely scattered Indian Ocean atolls that are presently attached to the crown colony of Seychelles for political administration. They are traditionally regarded as outside the archipelago geographically and are quite different from it in character, all being formed entirely of calcareous material, very low, and inhabited sparsely and irregularly.

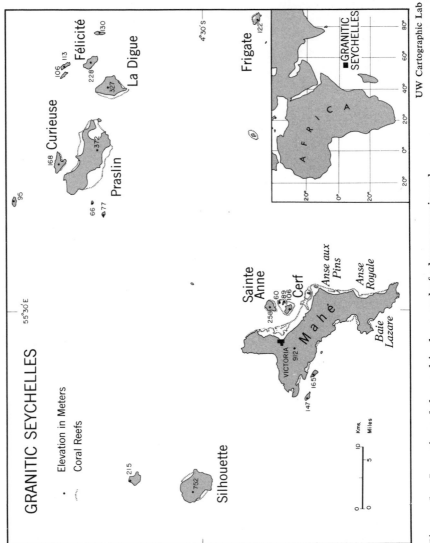

Figure 1. Location of the archipelago and of places mentioned

Whatever its early dislocations, the Seychelles mass must have been tectonically stable for a long time. Since being exposed to marine erosion, the granites and any overlying sedimentaries have been planed off and thinly mantled with fossil coral and debris to produce an extremely even platform at about 30 fathoms depth and about 200 miles across. At the time of its discovery, this bank was inhabited by dugongs and an abundance of sea turtles; it still teems with fish and sea-birds. When sea levels dropped during Pleistocene glaciations in high latitudes, the archipelago must have emerged as a single island with an area hundreds of times that of the present archipelago. The edges of the bank are precipitous and ocean depths of more than 1,000 fathoms separate it from other. enigmatic shoals on which the outlying atolls lie. The whole system of western Indian Ocean banks forms a great broken arc, concave toward Madagascar, cut off by abyssal depths from the continents and distinct from the mid-oceanic Carls-berg Ridge. The abyss east of the Seychelles seems to be a normal ocean floor, that to the west a peculiar geosyncline which has evidently been part of the Indian Ocean at least since the Triassic and probably since the Carboniferous (Heezen and Ewing, 1963; Baker, 1963; Browne, 1963; Matthews, 1965). The problematical geologic history of this region was one of the major concerns of the 1961–66 International Indian Ocean Expedition.

The present land area of the Seychelles is about ninety square miles, over half on Mahé. The islands are rugged for their size, the crystalline rocks forming gigantic boulder piles, towering cliffs, and domes (Plate 18). Coastlines are mostly bedrock or boulder slopes, locally called *glacis*, in many places a beautiful reddish color and strikingly carved by water (Plates 9, 20). Remains of fossil coral reefs, evidently marking an interglacial sea level about nine meters higher than at present, are attached beneath overhanging rocks on various islands (Gardiner, 1936; Baker, 1963; Russell and McIntire, 1965). Earlier reports of coral at higher levels (Horne, 1875; Keller, 1898) are now believed to be in error.

Where there are living fringing reefs, the coasts generally have white sand beaches backed by calcareous sand flats, locally called *plateaux* (Plate 13). Some of these extend inland for several hundred meters and, with a combined area of about five square miles, constitute the most extensive flat surfaces in the archipelago (Lionnet, 1952). Formed by steady accretion of beach material, with no dunes or storm ridges, their surfaces and bases are congruent throughout. Radiocarbon dating of reefal debris from the oldest, inner parts of the plateaux indicates that they began forming about three thousand years ago, when the postglacial rise of the sea had reached within about a meter of the present level and was slowing to a near stillstand (McIntire, *unpublished*). Plateau surfaces generally lie about three meters above low tide and about two meters higher behind the most exposed beaches. Tidal range is about one and one-half meters and buildup of beach ridges above that level is due simply to wave upwash. Complex tidal channels and swamps have formed where streams cut through the plateaux, mucky freshwater marshes where smaller streams are impounded.

The islands' best soils are on small alluvial deposits at the rear of the plateaux and in valleys behind. Residual upland soils weathered from the granites are generally lateritic, thin, and easily erodible. The alkaline plateau sands are black with organic matter near the surface but are low in some plant nutrients. Near the water table the calcareous sand is cemented into an impenetrable hardpan or platin, which outcrops as beach rock on eroding shores. This cemented layer forms within the fluctuating zone of the freshwater table, which generally lies about two meters below the surface in the wet season and slightly lower in the dry (Lionnet, 1952; Russell and McIntire, 1965).

The equatorial, oceanic climate has an unearthly mildness. At sea level, temperatures average around 27° C (80° F) all year, with absolute extremes varying not more than 5° C from the mean. Relative humidity seldom falls below 70 percent. The tropical cyclones (hurricanes or typhoons) that trouble higher latitudes are unknown, although they have passed near enough to the southward for the islands to be reached by heavy rollers generated by their winds. Fresh breezes and rough seas prevail in the season of the Southeast Trade, which generally sets in during April and blows steadily from May through October. This is the drier season, the sun usually shining most of the day through scattered cumulus clouds. Sometimes a week or two passes with no rain, but as a rule there are brief showers every two or three days. The Trades normally weaken in November and there are light, variable winds and calms until the Northwest Monsoon arrives, generally during December. Winds continue from the northwest, usually with gentle to moderate breeze force, until March, when variable winds and calms precede the re-advance of the Trades. During the Northwest Monsoon and transition seasons, sunshine can still be expected nearly half of the day but rains are comparatively frequent and heavy; most years have several days with 5 centimeters (2 inches) or more of rain; nearly 50 centimeters (19 inches) fell on Victoria one February day in 1958. Records for ten years or longer are available from eight coastal stations on Mahé, where rainfall has averaged between 175 and 265 centimeters (70 and 105 inches) a year; two interior stations at about 300 meters elevation on Mahé have recorded averages of about 275 centimeters (110 inches); and at two coastal stations on Praslin and La Digue, rainfall has averaged about the same as at the drier Mahé coastal stations. Records from Félicité, Curieuse, and Silhouette cover too short an interval to give good averages but they fall within the range of the Mahé coastal stations for the corresponding years. Contrasts in average rainfall between different coasts are evidently quite moderate, as is most year-to-year variation, although a given place may get twice as much rain in some years as in others (Colonial Office, Great Britain, 1949–65; East African Meteorological Department, 1956–61, 1960–62; Lionnet, 1957–61; W. Kuepper, *personal communication*).

Thus, even with infertile soils, the climate favors a lush

Plate 1. Coconut plantation on calcareous sand plateau, Anse aux Pins, Mahé

Plate 2. Jungle of volunteer cinnamon and planted coconuts on granite hills behind Anse aux Pins, Mahé

vegetation (Plate 6) and the islands were originally largely covered with broadleaf evergreen rainforests. In favored places these were dominated by magnificent palms and great hardwood trees of the sapote, dipterocarp, and other families. The most abundant trees were smaller species of palms and screwpines, forming the lower story in tall forests and dominating dry rocky sites (Schimper, Schenck, and Diels, 1922; Vesey-Fitzgerald, 1940; Dayer, 1962; Jeffrey, 1962, 1963). Beneath its luxuriant facade, the aboriginal inland vegetation was extremely poor floristically, including only about 80 species of ferns and about 170 of flowering plants. Approximately half of these species were widespread outside the archipelago, including most of the ferns, which have wind-borne spores, and many cosmopolitan marsh plants with seeds perhaps dispersed by migratory birds. About a dozen of the ferns and nearly a hundred of the flowering plants were endemic, differing from their nearest relatives in the outside world in varying degree. Some show only slight divergence; others belong to peculiar genera; one flowering plant, supposed to have become extinct about fifty years ago, was the only known member of a family, the Medusagynaceae, distantly related to the camellias. Geographic affinities of both cosmopolitan and endemic species were diffuse, substantial groups having their closest relations in Africa, Asia, Madagascar, and other paleotropic islands (Christensen, 1912; Summerhayes, 1931).

The native land fauna, apart from numerous insects and other invertebrates, was also exceedingly poor in number of species. In addition to wide-ranging pelagic species and shorebirds, there were about a dozen species of endemic land birds. The only native mammals were fruit bats, or flying foxes, and the only large terrestrial animals of any sort were alligators and giant tortoises (Frappas, 1820; Gardiner, 1936; Loustau-Lalanne, 1961).

Arrival of the Seychelles biota by overland migration on a now sunken or dismembered continent was visualized in the classic Limuria and Gondwanaland hypotheses and still finds active support (Steenis, 1962; Jeffrey, 1964). Immigration entirely by transoceanic dispersal was suggested as an alternative long ago (Engler, 1882). Conflicting opinions were held by members of the 1905 *Sealark* expedition to the region, the entomologist (Scott, 1933) arguing that land bridges were needed to account for insect and plant distributions, the expedition leader (Gardiner, 1936) concluding that the whole biota arrived by long-range dispersal. Gondwanaland persists as a possibility in modern geologic theory but in a form that does not supply overland migration routes to the Seychelles; evidently any such "continent" was always divided by geosynclinal seaways and, in any case, was dismembered at least 200 million years ago, long before the origin of the flowering plants (Baker, 1963; Fairbridge, 1965).

It seems to me that plants and animals could have reached the Seychelles by transoceanic dispersal as readily as they did many other islands, such as the Hawaiian group, for which continental connections are inconceivable. The small number of species, their diffuse geographic affinities, and their varying grades of divergence from outside relatives are exactly what might be expected from gradual accumulation of long-range migrants, some arriving by freakish and unrepeated dispersal, others reintroduced occasionally or regularly. This question will be further explored in the section on natural dispersal mechanisms of the aboriginal coastal species.

History 2
of
Human
Intervention

Entry of the Seychelles into recorded history was curiously late, considering the antiquity of Indian Ocean seafaring. Clemesha (1943) and Villiers (1952) have suggested that the Indian Ocean was the birthplace of deep-water sailing and shipbuilding. Certainly coastal navigation between Red Sea, Persian Gulf, and Indian ports began in antiquity. By the first century B.C., Greek ships were crossing open sea directly from the Gulf of Aden to Malabar. In the early centuries of the Christian era, Greek and Roman trade with India sometimes involved over a hundred ships a year, while other vessels were coasting East Africa to the vicinity of Zanzibar. During medieval times, Persian, Indian, and Arab ships ranged all the way from East Africa and Madagascar to China (Hourani, 1951; Mookerji, 1957; Wheatley, 1964).

Several waves of Malayan and other East Indian immigrants reached Madagascar by mysterious prehistoric pathways (Hornell, 1934; Grottanelli, 1955; Boiteau, 1958; Hänel, 1959). Historic Chinese expeditions through the western Indian

Ocean reportedly reached East Africa between 1417 and 1433 (Freeman-Grenville, 1962; Wheatley, 1964).

Arab knowledge of the Seychelles has been suggested on the basis of imaginative old charts and legendary travelers' tales of the kind that evolved into the Sinbad stories (Dumont d'Urville, 1834; Belcher, 1843; Bulpin, 1960; McEwen, 1961). Archaeologic evidence of pre-European contacts is lacking, cave excavations in the islands having been fruitless (Gardiner, 1936), and historic records in the outside world give no clear evidence of such contacts. Paul Wheatley (*personal communication*) wrote: "I have often reflected that in the pre-European historic period the Seychelles were remarkably unnoticed. I don't remember ever having come across a recognizable reference to them in the rutters of the Chinese or Arab pilots, some of whom left rather detailed accounts of their voyages in the Indian Ocean."

Plate 3.   Estate-worker's house in coconut plantation near Police Point, Mahé

Plate 4. Cinnamon oil distillery surrounded by spent leaves, near Grande Anse, Mahé

Early voyagers would have usually passed far to the north and west of the Seychelles, both by preference for coastal navigation and because the monsoon wind patterns strongly favor such an arc. Coming up the east African coast on Vasco da Gama's first voyage, the Portuguese found four Indian ships in the harbor at Malindi and recruited a Gujarati pilot who impressed them greatly as a navigator. From Malindi he guided them up the coast for a few days before standing eastward, well to the north of the Seychelles. They saw no land until sighting the Ghats 23 days after leaving Africa (Clarke, 1803; Ravenstein, ed. and transl., 1898, da Gama). The Portuguese maintained a long-standing bias in favor of this route, but occasional shortcuts were essayed. Antonio Galvão, who went to India in 1527 and later became governor of the Moluccas, credited da Gama with a direct crossing from Mozambique to India on his second voyage in 1502; on this crossing he discovered an island at latitude 4° S which he named Almirante (Hakluyt, transl., 1862, Galvão). Other great Portuguese navigators, João da Nova, Affonso d'Albuquerque, Tristão da Cunha, and Pedro Mascarenhas, were exploring beyond Madagascar early in the sixteenth century. A superfluity of islands appeared on contemporary charts of the western Indian Ocean, shown so diagrammatically and with such varying longitudes that their identities are undeterminable. If sightings of the Seychelles contributed to these, they must appear under such cartographically wandering names as Almirantes, Mascarenhas, Seven Sisters, Three Brothers, and variants in different languages (Frappas, 1820; Pridham, 1846; Fauvel, 1909; Keller, 1898; McEwen, 1961; Scott, 1961). No matter how likely Portuguese sighting of the Seychelles may be, there is no reason to assume landings. Scott (1961) notes that the Portuguese were so well equipped for long voyages that they were indifferent to isolated islands and disinclined to use any of them except perhaps as landmarks.

Seychelles history begins on January 19, 1609, when a British East Indiaman, the *Ascension*, commanded by Alexander Sharpeigh, happened onto a group of islands taken to be the Almirantes of old charts. The *Ascension* anchored for a week in what is now called Victoria Harbour before going on to become the first British ship to sail the Red Sea. Separate journals of the two factors, John Jourdain and William Revett, and of the boatswain, Thomas Jones, give a delightful, circumstantial account of the islands, which they saw as an earthly paradise, "a very good refreshing place for wood, water, coker nutts, fish and fowle, without any feare or danger, except the allagartes; for you cannot discerne that ever any people had been there before us." The journals agree on the tameness of the birds, the abundance of great tortoises and coconuts, and the magnificence of the forests, "as good tymber as I ever sawe of length and bignes, and a very firme tymber. You shall have many trees of 60 and 70 feete without sprigge except at the topp, very bigge and straight as an arrow." This episode was forgotten for almost three hundred years until Admiral Sir William Wharton, who had carried out hydrographic surveys in the Seychelles, identified the islands described in these journals with their present names (Foster, ed., 1905, Jourdain; McEwen, 1961). After the *Ascension* left, the islands were undisturbed by man for over a century, unless there is substance in popular legends of pirate visits (Pridham, 1846; Tonnet, 1906; Mackay, 1941; Ommanney, 1952), which still inspire amateur excavations.

Effective discovery, leading eventually to colonization, came in the mid-eighteenth century with a burst of exploration by the French East India Company, directed by the great governor of Mauritius, Bertrand François Mahé de la Bourdonnais. In November of 1742, two Mauritian ships, the *Elisabeth* and the *Charle*, effected a landing on the southwest coast of Mahé, which they took for one of the Three Brothers, at a bay that still bears the Christian name of one of the captains, Lazare Picault. The journals of Picault and the other captain, Jean Grossin, report abundant tortoises, fine timber suitable for great masts, and coconuts, which they loaded by the hundreds from palms lining the beach. They had gone ashore armed to fight but found no sign of any previous visitors (Grant, 1801; Frappas, 1820; Fauvel, 1891, 1909). The chronometer and accurate longitude reckoning were yet to be developed, and these explorers have been charged with an incredible navigational error of three hundred leagues on this voyage (Après de Mannevillette, 1775). However, Picault had no trouble returning two years later to take formal possession of the archipelago, which he named Isles de la Bourdonnais. He named the main island Mahé and two others, presumably Praslin and Curieuse, respectively, Isle de la Palme "because it bears abundant palmistes" and Isle Rouge "because in the middle there is a mountain whose earth is red" (Fauvel, 1909). The latter observation raises a question about modern indictment (Gibson, 1938; Swabey, 1961) of the proprietors and inhabitants of Curieuse for vandalistic deforestation leading to catastrophic erosion of the uplands (Plate 20).

In 1756 the next visiting Mauritian ship, the *Cerf* under Captain Corneille Nicholas de Morphey, gave its name to another island and renamed the group after a French politician, the Vicomte Moreau de Séchelles. Morphey noted that the southwest coast of Mahé was fringed with coconut palms which he believed had grown from nuts cast up by the sea, as the palms were not found more than twenty paces from the shore; the great bay on Mahé's northeast side was screened with impenetrable mangroves (Fauvel, 1909).

Exploration was suspended during the Seven Years War and the shift of Mauritius from East India Company control to direct rule by the French crown. Finally in 1768 the intendant of Mauritius, Pierre Poivre, initiated a series of expeditions that led directly to settlement. The Chevalier Marion du Fresne, commander of the 1768 expedition, named two islands after his ships, *La Digue* and *La Curieuse*, and another after a French minister, Gabriel de Choiseul, Duc de Praslin, but reported the last island to be impenetrable and uninhabitable (Fauvel, 1909). However, enough penetration was effected to uncover a mystery of long standing — the hitherto unknown source of what had been considered since medieval times to be one of the wonders of the world.

Here at last were living groves of the celebrated coco de mer, double coconut, or Maldive nut. This great palm, *Lodoicea maldivica*, is neither a coconut nor a native of the Maldives, being endemic to the interior of Praslin and Curieuse. Its nuts, the largest and most bizarre of all seeds, had occasionally drifted to the Maldives. They were widely traded in Asia and Europe as a fabulously valuable medical panacea and infallible antidote to poisons (Parkinson, 1640; Rumpf, 1741–50; Frappas, 1820; Seemann, 1863; Gray, ed. and transl., 1887, Pyrard; Burkill, 1935; Bailey and Vesey-Fitzgerald, 1942; Jeffrey, 1964). Parkinson recounted legends that the tree grew beneath the sea or came from an island called Palloyes "seen by some that looke not for it but cannot be found by them that seeke it." He was scornful of expeditions which were sent by Maldive kings to search for the island and returned empty-handed and affrighted; he predicted that the land where the trees grew could be found by stout not timorous, religious not superstitious, not weaklings or fools, but judicious and industrious men. Duchemin, captain of *La Digue*, stoutly shipped a whole cargo for sale in India. Soon other French and British sailors, more industrious than judicious, glutted the market with nuts they expected to sell for kings' ransoms. When Poivre received a living coco de mer in Mauritius, he immediately asked for a whole shipload, "enough to plant the whole of our Tonneliers Island and the surroundings of our principal ports" (Fauvel, 1909).

This sensation unfortunately monopolized the attention of the first scientific visitors to the Seychelles. The Abbé Alexis Marie Rochon, geographer and academician, was in the islands for a month in 1769 and the naturalist Pierre Sonnerat passed through in 1771 (Rochon, 1791; Sonnerat, 1776). In their preoccupation with the coco de mer, they lost the last chance to record the primeval state of the archipelago.

COLONIZATION
AND EARLY PLANTATION PERIOD

In 1770, the *Télémaque* arrived from Mauritius with the first colonists, under command of de Launay. This group, about thirty people including slaves, landed on Sainte Anne Island. The next year the *Etoile du Matin* landed about forty more people from Réunion, under the command of Gillot, on Mahé. The whole venture was originally a semiprivate scheme, sanctioned by Poivre but financed by a colorful Breton promoter, Brayer du Barré, who attempted absentee management of his concession from Mauritius. His grandiose and diffuse plans involved exploitation of coco de mer, coconut oil and fiber, timber, and marine products, as well as plantations of innumerable subsistence and commercial crops. In 1771 and 1772 he glowingly reported successful coffee, cotton, and tobacco planting, abundant harvest of rice, sweet potatoes, maize, manioc, and various vegetables. Shortly after, he claimed establishment of gardens with many kinds of fruit

Plate 5. Beach morning-glory carpeting abandoned clearing, planted coconuts and *Casuarina* grove in background, near North Point, Mahé

trees, sugarcane, and plants producing oils, gums, dyes, and spices. For many years Poivre had been trying to break the Dutch spice monopoly by establishing spice plantations in Mauritius. When Gillot visited Mauritius in 1772, Poivre entrusted him with some of the precious plants to take back to Mahé, where they could be grown secretly and in safety from the hurricanes that trouble Mauritius. Gillot was charged with establishing a sort of branch of the famous Mauritius Botanic Gardens, a royal garden independent of Brayer du Barré's enterprise. A map from about 1772 shows Gillot's *Jardin du Roi* as four neat squares on the shore of Anse Royale with young clove, nutmeg, pepper, and cinnamon trees shaded by banana plants (Fauvel, 1909; Ly-Tio-Fane, 1958; Louis, 1961).

All these pretty plans were soon blighted. Before the end of 1772 Poivre was replaced by a new intendant who had little interest in the Seychelles and recalled most of the colonists to Mauritius. Brayer du Barré was exiled from Mauritius after a scandal over promotion of an imaginary Seychelles silver mine. He disappeared from local history with a parting shot at de Launay for letting the fruit trees, sent at great expense, die at the edge of the sea while everybody was off gathering shells. Passing English ships repeatedly raided Praslin and Curieuse, taking coco de mer and setting fire to the forests. Gillot and de Launay feuded, while the slaves revolted and ran away. Two distinguished visitors, Jean François de Galaup, Comte de la Pérouse in 1773 and Louis Antoine de Bougainville in 1775, found the colony in chaos, the colonists hunting turtles instead of cultivating crops, perishing from hunger and at each others' throats, deserters from the garrison burning the forests and stealing slaves, runaway slaves ravaging Gillot's *Jardin du Roi*. In 1777 the slaves allotted to care for this garden were taken away and the few remaining spice trees abandoned to Gillot as his private property; he tended them with pathetic devotion until 1780, when the commandant had them destroyed to prevent capture by an imaginary English raiding vessel (Fauvel, 1909; Ly-Tio-Fane, 1958; Louis, 1961). Although neither Gillot's nor later efforts ever led to successful spice plantations, they did eventually have important consequences, mainly through the escape and naturalization of cinnamon.

In 1786 another visitor, Saulx de Rosnevet, reported the colony still in a precarious state. The Mahé and Sainte Anne establishments consisted of a tiny garrison, four French habitants and one freed slave, together holding a total of about a hundred slaves. A single habitant with a dozen slaves was on Praslin. The colony had no roads and a total fleet of five pirogues (Plate 7) for communications. Of the modest 120 arpents (125 acres) claimed by the habitants, only about half were under cultivation. Rosnevet doubted that the colony could support more people, the soil being so poor that after a few harvests it returned scarcely twice the maize or rice seed planted. He was more hopeful about timber exploitation and also noted that a herd of about sixty cattle had been running wild on Mahé for two years and was increasing (Fauvel, 1909).

At this nadir, in 1786, the colony fortunately received an admirable new commandant, Jean Baptiste Philogène de

Malavois, engineer, agronomist, and geographer. Malavois rebuilt the rundown defense and administrative establishments and set about regulating land occupation, which had been chaotic and shifting at the convenience of the habitants. One problem was encroachment on the royal coastal reserve, which the Mauritian authorities charged him with maintaining (Fauvel, 1909).

This reserve, known as the *Cinquante Pas du Roi* or *Pas Géométriques*, was prescribed by general French colonial policy. With minor changes during the seventeenth and eighteenth centuries, it was designated as a continuous strip extending about a hundred meters inland from high-tide mark or from where vegetation began. Except for batteries, fortifications, and access routes, the strip was supposed to be left uncleared as an obstacle to attack and as a source of timber needed by visiting ships. Private occupation and exploitation were proscribed, although artisans and fishermen might be allowed to lodge there if denied a place by proprietors of inland concessions (Rougon, 1876). In Mauritius this institution has survived all political changes and remains strikingly evident in the landscape as a tree belt under government control between the sugar estates and the sea (Sauer, 1961).

In the Seychelles, any success Malavois had in reserving this land was ephemeral. Coconuts, going to waste from the wild palms fringing the beaches, were inviting trespass. In Malavois' time these royal coconuts were supplying about 6,500 liters of oil a year for local use. Malavois suggested a legal, joint harvest every three months, the fruits to be divided in the commandant's presence with a share to the government (Fauvel, 1909; Dayer, 1962), but there is no record that this was ever done.

Malavois did institute orderly land tenure, conceding to each habitant an area of 108 arpents (112 acres) fronting on the coast and granting areas one-fourth as large to freed slaves. This policy has left a lasting mark on the land-holding pattern, even if blurred by changes that will be mentioned below.

Malavois thus reserved most of the land area for concession to the many future colonists included in his reckoning. He was mildly optimistic about agricultural potential of the islands, provided suitable crops and techniques were used. He approved of the habitants' wet rice cultivation in coastal marshes but not their planting of dry rice and maize in the uplands. He recommended that these be restricted to good valley soils, with terracing and manuring, and that ground-covering sweet potatoes and manioc be substituted on quickly exhausted upland soils. He noted that bananas, sugarcane, cotton, tobacco, and pineapple grew well, and deplored the indifference of the people to vegetables introduced from Mauritius and their neglect of fruit-tree propagation. Malavois also wrote a series of detailed memoirs on the colony's timber resources, noting that the lushness of the forests had led to exaggerated ideas of the timber content; the bulk of the forests was composed of palms and screwpines; the fine timber trees were not rare but uncommon. He proposed bringing in skilled sawyers, controlled cutting for export to Mauritius and India, and replanting cleared land with selected species. The population of the colony increased to 22 habitants with several hundred slaves during the six years of Malavois' administration (Fauvel, 1909; Dayer, 1962).

In 1792 the French Revolution arrived in the form of a ship bearing citizen administrators, who proclaimed liberation of the slaves and universal suffrage. This shock was not very profound or enduring, normal conditions returning with the appointment in 1794 of the next commandant, the Chevalier Jean Baptiste Quéau de Quinssy (Dayer, 1963). This shrewd and adaptable aristocrat was to remain serenely in charge of the colony through all the overturning of the French and Mauritian governments and the Napoleonic wars.

Meanwhile, the Seychelles developed a thriving shipbuilding and shipping industry and a diversified plantation system, along lines proposed by Malavois, based on labor of slaves from Africa and Madagascar. At first most of the planted

area was in food crops for home consumption and for vic-
tualing ships. Rice was grown in coastal marshes, with dry-
land rice, maize, millet, manioc, sweet potatoes, and bananas
in the uplands. There were small plantings of pineapples,
lemons, tobacco, coffee, pepper, cloves, and enough sugarcane
for local rum distillation. The abundant protein from fishing
and turtling was augmented by many pigs and fowl and a
few hundred cattle and sheep (Frappas, 1820; Unienville,
1838; Webb, 1960a, 1962; Lionnet, 1962).

Until 1800 exports were almost entirely natural products,
at first mostly tortoises and sea turtles, of which over 10,000
are believed to have been shipped between 1773 and 1800;
turtle shell continued to be a major export thereafter. No
longer highly valuable, coco de mer was being exported in
small quantities as a curiosity, just as it is today. Salt fish, fish
oil, coconut oil, and timber also appear prominently in cargo
lists before 1800. The first recorded export crop was cotton,
appearing in the lists in 1796 and becoming the main export
after 1802. By 1810 the islands had nearly 3,000 acres in cotton,
roughly equal to the area in provision crops; the population
was about 3,500, including over 3,000 slaves (Frappas, 1820;
Unienville, 1838; Toussaint, 1965).

With the blockade and fall of Mauritius in 1810, many
planters moved to the Seychelles with their slaves in the hope
of escaping British control. De Quinssy occasionally capitu-
lated to passing British men-of-war and by the Treaty of
Paris in 1814 the Seychelles became a British possession. This
formal change of ownership had curiously little immediate
effect on the colony, de Quinssy remaining in office for another decade, French laws and land titles continuing to be
recognized, and cotton-planting being carried on as before
(Webb, 1960b).

By 1819 Frappas found most of Mahé cleared of its original
forest; he doubted that cropping of the thin, quickly eroding
upland soils would last long. At that time the future town of
Victoria, still called merely *l'établissement*, consisted of a few

Plate 6. Pirogue sheds on sheltered bay, a few mangroves at right, planted coconuts and
dooryard trees on steep slope, Cascade,
Mahé

wooden houses. Population of the colony reached a tempo-
rary peak in 1830 at about 10,000, including over 9,000 slaves.
A map at that time shows virtually the whole littoral of Mahé
and Sainte Anne divided into land grants, usually of the 108
arpents prescribed by Malavois. Settlers were also established
on Praslin, La Digue, and Silhouette, other islands being
largely conceded but not yet occupied. Cotton-planting then

declined in the face of American competition and by 1831 had taken second place to sugar-planting. An exodus of settlers with their slaves ensued and by 1840, five years after the abolition of slavery, the total population was down to about 5,400, less than 1,000 of whom were on islands other than Mahé (Frappas, 1820; Laplace, 1833–39; Martin, 1836; Unienville, 1838; Pridham, 1846; Pike, 1873; Fauvel, 1891; Webb, 1960*b*, 1962).

RISE OF COCONUT PLANTATIONS

With emancipation, the old plantation system ended abruptly, the former slaves refusing fairly unanimously to continue in their old jobs for wages (Belcher, 1843). According to local tradition, the planters shifted directly to coconuts as a solution to the labor shortage, coconut plantations requiring only about one employee per ten acres.

The local chronology is amazingly close to that of the world coconut industry. Coconut oil and copra first became important in international trade in the 1840's, production coming initially from natural groves and small plantings of the sort made for domestic use since antiquity. Large commercial plantations were started in Ceylon in 1842; the earliest large-scale production developed there and in Malaysia in the 1860's (Pieris, 1961; Child, 1964).

In the Seychelles, coconut-planting had been advocated from the outset by du Barré, Gillot, and Malavois, and a few plantings may have been started quite early (Martin, 1836; Fauvel, 1909; Dayer, 1962). As late as 1819, coconuts were still reported as strictly seashore trees and were not mentioned among the crops (Frappas, 1820). By 1840 the hábitants claimed that they already had established plantations capable of producing several hundred thousand liters of oil annually, but were being held back by shortage of labor, and they were seeking permission to introduce Indian labor (Pridham, 1846).

Plate 7. Pirogue on cove beach, edge of littoral thicket overhung
by coconuts, Anse Royale, Mahé

This estimated yield is over a hundred times the early production from wild seashore palms; even if exaggerated, it
indicates that planting had begun before emancipation, as it
takes a coconut about seven years to begin bearing. In 1842
the island of Sainte Anne was described as a forest of coconuts,
although cotton and coffee were still listed ahead of coconuts
in the colony's commercial products (Belcher, 1843).

By 1850 coconut plantations were furnishing considerable
oil, a few estates were still producing sugar, and there were
only small, unimportant plantings of cacao, coffee, cloves,
rice, and tobacco (Keate, 1850). By the early 1860's coconut
oil, shipped to Mauritius, had become almost the sole export.
Relics of cane, tobacco, and coffee plantations were maintained mainly for domestic needs. Bananas, rice, maize, sweet
potatoes, and manioc continued as provision crops (Pelly,
1865; Kersten, 1871; Pike, 1873). Unlike Mauritius, the Sey-

chelles never brought in a large Indian labor force, but between 1861 and 1872 nearly 2,500 slaves liberated from Arab dhows were landed by British vessels and apprenticed to the coconut-planters (Webb, 1960*b*).

In 1871 and 1874 the sub-director of the Mauritius Botanic Gardens visited all the main Seychelles islands and reported coconut plantations being extended annually, older ones occupying the coastal plateaux and newer ones reaching an elevation of 1,500 feet on the mountain slopes. Even Félicité, a remote little island entirely crown land, had been leased for coconut-planting and mostly cleared (Horne, 1875). After a lifetime of travels painting landscapes around the world, now displayed in a special building at Kew Gardens, Marianne North (1892) remembered the Seychelles in 1883 as having coconut palms mounting higher up the mountains than any other place she had seen. All accounts toward the end of the century agree that coconut plantations blanketed the coastal

Plate 8. Littoral thicket and planted coconuts on densely inhabited shore (Fig. 7), beach rock on left exposed at low tide, Anse aux Couches, Mahé

plateaux and lower mountain slopes (Fauvel, 1891; Brauer, 1896; Keller, 1898; Schimper et al., 1922).

By 1900 the species had thus become almost the sole support of a population totaling nearly 20,000 and growing rapidly by natural increase (Webb, 1960b). By then the *Pas Géométriques*, which included a large part of the best coconut lands, had been forgotten as an archaic edict of a remote administration. As a colony of a colony, the Seychelles had been left largely to their own devices, sometimes without regular external communications. When they were finally removed from this limbo and made a separate crown colony in 1903, the new Secretary of State questioned the lack of any reference to the *Pas Géométriques* in an ordinance being drafted to regulate crown lands. He was informed by the local court that there was no law in force or applicable, and that ". . . in point of fact there does not appear to be any land respected as *Pas Géométriques*. Owners occupy up to the beach" (Anonymous, 1903). The reserve has since been regarded as a dead issue.

The economic supremacy of the coconut has continued unshaken to the present, but there have been some changes in processing the crop. Early in the present century, copra, previously a minor product, replaced coconut oil as the main export. For a long time copra was mostly shipped to Europe; since 1951 it has gone almost entirely to India, where it sells at a premium because of its quality. The copra is extracted from the nuts and dried on the plantations, either in the sun or in ovens fired by the superabundant shells and husks. Only a minute fraction of the husks are used by the local coir fiber industry, which has functioned sporadically for many years; however, a newly established coir factory produced 400 tons of fiber for export in 1965. Substandard copra is pressed for domestic oil needs, the poonac residue serving as livestock feed. A fair number of fresh nuts are diverted to food and drink, the average citizen consuming an estimated 150 nuts a year. However, nearly 90 percent of the annual harvest of from 40 to 50 million nuts is converted to export copra. Recorded exports, presumably including some produce of the outlying islands, which have about one-quarter of the coconut acreage of the granitic group, climbed from about 1,500 tons per year prior to 1910 to about 5,000 tons in the late 1920's; they have fluctuated between 5,000 and 7,000 tons ever since. Copra constitutes over three-quarters of the colony's modest exports, with an annual value of about 6 million rupees (slightly over 1 million dollars). Expansion of coconut acreage is inconceivable, as plantations have long filled all suitable sites and been pushed far beyond. About three-quarters of the yield comes from about one-quarter of the palms, mainly on the coastal sand flats. Marginal upland plant-ings are being abandoned rather than extended (Colonial Office, Great Britain, 1949–65; Durocher-Yvon, 1953; Cooke, 1958; Lionnet, 1957–61, 1962, *personal communication*; Rowe, 1959; Webb, 1960*a*).

MODERN HISTORY OF SUBSIDIARY

COMMERCIAL CROPS

Even though copra has been continuingly profitable, concern over its static prospects and over the vulnerability of any monoculture has led to innumerable attempts by planters and government to diversify the agricultural base. Of all the exotic plants tried, only three have been commercially suc-cessful during the last hundred years: vanilla, patchouli, and cinnamon.

Vanilla, a terrestrial orchid native to tropical America, was introduced to the Seychelles in the eighteenth century (Anon-ymous, 1800) and was noted as a minor product in the early 1860's (Pelly, 1865; Kersten, 1871). A separate introduction via Réunion, perhaps cuttings of a superior variety, came in 1866 and numerous plantings were laid out by 1875. Train-ing the vines on supporting poles or trees, Seychelles producers developed skill in the complex arts required for vanilla cultivation, flower induction, pollination, and curing. During the late nineteenth century the Seychelles produced more vanilla than all other British colonies combined. This was never a major share of the world crop but it has been important for hundreds of small growers on Mahé, Praslin, and La Digue, and was the basis of some local fortunes. Troubles, starting in 1904, with drought, disease, and competition from synthetic vanillin almost destroyed the local industry. Planting has been expanding again since 1956; the crop is now worth several hundred thousand rupees in good years (Horne, 1875; Dupont, 1938; Colonial Office, Great Britain, 1949–65; Lionnet, 1957–61, 1958–59, 1962).

Patchouli, an East Indian herb of the mint family yielding an essential oil with a long history in perfumery, was introduced to the Seychelles from Mauritius late in the nineteenth century. The oil was first exported in the 1920's; a great boom came during World War II, when prices shot up a hundredfold after the usual Malaysian supply was shut off. For a time, patchouli was second to copra in export value, being grown by large estate-owners and small holders alike. Profits and acreage declined drastically after the war. The crop is now hardly worth mention, exports being valued at a few thousand rupees a year (Dupont, 1927; Lionnet, 1957–61, 1962; Rowe, 1959).

During most of the years since 1908, cinnamon has been the main export after copra. Its history is curious because it was introduced early and deliberately, but exploitation was long delayed and involves feral rather than cultivated stands. Of the four kinds of spices in Gillot's *Jardin du Roi*, only cinnamon survived the garden's destruction, according to Malavois (Fauvel, 1909). Spread widely by birds, the species soon became naturalized (Pridham, 1846). On Mahé and Silhouette, by the late nineteenth century it commonly dominated secondary forest vegetation and volunteered in coconut plantations. By an underestimate of Poivre, the species was long thought to be a worthless kind, a mistake that was not corrected by the report of a visiting botanical expert (Horne, 1875). Some cinnamon was exported in the early nineteenth century (Pridham, 1846) and a small factory was distilling cinnamon oil at Cascade on Mahé before 1900 (Guérard, 1891), but mostly the plants were left alone to multiply and grow into trees.

Finally, in 1902 the local cinnamon trees were identified by the newly established Department of Agriculture as belonging to the best species of Ceylon cinnamon, and intensive exploitation began. Bark-stripping reached a peak in 1908; big trees were ruthlessly felled, in which case they usually resprouted, or were barked standing, in which case they died. After 1915, exploitation shifted by necessity to distillation of leaf oil from sprouts and seedlings too small to yield good bark. Firewood cutting for the distilleries deforested mountain areas, and the distilleries were forced to concentrate along the coast where abundant coconut trash was available as fuel. About fifty distilleries are now operating, each surrounded by great drifts of spent cinnamon leaves (Plate 4), some of which are used as fuel or spread as a fragrant mulch on nearby coconut plantations.

The vitality of this feral cinnamon is amazing. In accessible areas sprouts are commonly slashed for leaf-plucking about every one and one-half years, but the coppice continues to grow back and produce leaves at the rate of nearly a ton per acre per year. In more remote areas cropping is sometimes delayed for five years or more, regrowth of sprouts then being sufficient for profitable bark harvest. Bark production has recovered strongly since 1942, again exceeding leaf oil in value and reaching an all-time record of nearly 1,900 tons in 1965. Together, cinnamon bark and leaf oil now constitute an export valued at about 1.5 million rupees a year (Dupont, 1911–22; Gibson, 1938; Colonial Office, Great Britain, 1949–65; Lionnet, 1957–61, 1961, 1962, *personal communication*; Webb, 1960a).

Plate 9. Screwpine emerging from crevice and beach morning-glory sprawling on granite, tip of exposed promontory, Pointe au Sel, Mahé

Plate 10. Planted *Casuarina* grove in foreground, natural thicket of screwpine at right center, looking toward Anse aux Couches from Pointe au Sel, Mahé

Only small areas of the Seychelles are now planted to subsistence crops, mainly in dooryard gardens and tiny patches under the coconuts. The islands depend on large imports of rice and other food staples exchanged for copra and cinnamon exports. The 1960 agricultural census enumerated 17,000 acres of coconut plantations in the granitic islands, the Seychelles proper, and the actual total was probably about 20,000 acres. There were then about 700 acres of vanilla, mostly under coconuts. About 14,000 acres of cinnamon were enumerated, half under coconuts (Plate 2) and half in pure stands in the mountains above the coconut belt (Webb, 1960a). Cinnamon has been tolerated in some plateau coconut plantations in the belief that it may function as a nutrient pumper, but current opinion is that it is competitive with coconuts, so it has been cleared from most of the better lands. Alone or together, coconut and cinnamon occupy about half the islands' surface. The remaining lands are generally very rough areas of scrub or bare rock.

Very little primary forest remains, with the glorious exception of some fine coco de mer reserves on Praslin. Extensive mountain areas degraded by cutting and burning have reverted to the government. Attempts have been made at their reforestation, and there has been no lack of official reports by visiting experts recommending forest conservation (Horne, 1875; Gibson, 1938; Swabey, 1961; Jeffrey, 1963). Laws providing for mountain and river reserves, on both private and crown lands, have theoretically been in force for nearly a hundred years, but shifts in policy and lack of staff have hindered consistent execution of plans and it is doubtful how much could have been done with these difficult lands at best. At present, the highlands are largely unproductive and uninhabited, while human activities and settlement are concentrated along the coastal fringes.

The total population of the granitic islands was slightly over 40,000 at the 1960 census and was expected to reach 45,000 by 1966. About 80 percent of the people live on Mahé and most of the rest on Praslin and La Digue (Webb, 1960*b*). Victoria is a town of 10,000, the governmental and business center and port through which all overseas commerce goes. There is nothing larger than a hamlet elsewhere in the islands. Rural settlement is scattered through coastal sand flats and rocky coves. Some estate buildings are imposing but the ordinary houses are inconspicuous; many are simply thatched huts tucked under the palms (Plate 3). Innumerable pirogue sheds dot the coast, fishing being more important as a food source than livestock or poultry. Some estates run small herds of cattle but the islands are not heavily grazed.

Plate 11.   Beach morning-glory retreating with onset of Southeast Trade season, waves at high tide reaching edge of littoral thicket (Fig. 3), Police Bay, Mahé

Consolidation and fractionation of holdings have intruded into Malavois' scheme of standard-sized land grants. There are hundreds of small holdings, often in complex multiple ownership and lease arrangements; many are too small to support a household by agriculture alone. Government resettlement schemes have added a new pattern in a few localities by allotting small plots for subsistence and market gardening. Generally these are carved out of crown lands by terracing hill slopes, but one new project, expected to accommodate eighty families by the end of 1966, is located on the coastal plateau of Praslin.

The old pattern of large estates remains the general rule, about three-quarters of the colony's land being in holdings of one hundred acres or more (Rowe, 1959; Webb, 1960a). The old French colonial families are still mainly in possession, but are being joined as large landowners by members of the small Indian community who, together with a few Chinese, conduct most of the colony's mercantile and commercial enterprises. The comparatively few British residents are generally in government or retired from colonial service. The general population is derived from a complex racial mixture, predominantly African but with strong Madagascan, Asiatic, and European elements. The folk culture is heavily influenced by French traditions, and the Creole language has developed primarily out of French.

Against this background, the effects of historic and present human activities on the coastal vegetation will be considered after an attempt to reconstruct its aboriginal state.

Aboriginal 3
Coastal
Vegetation

Coastal vegetation, as here defined, occupies habitats where microclimatic and edaphic conditions are strongly controlled by the nearby sea: tidal flats, spray-swept rock, beaches and coastal sands without mature soil profiles. There is usually a sharp transition in vegetational physiognomy and species composition between this pioneer fringe and the ordinary inland vegetation growing on stable substrates.

POSTULATED COMPONENT SPECIES

A total of 57 species are here suggested as ancient members of the Seychelles coastal vegetation (see Appendix 1 for list, and explanation of symbols used below). The list is partly supported by direct historical evidence.

The coconut (*Cocos nucifera*), appearing repeatedly in early explorers' accounts, was the only species clearly recorded prior to colonization. In 1756, the *Cerf's* Captain Morphey reported mangroves, without distinguishing species, and also gave ambiguous descriptions of two coastal trees which suggest *Calophyllum* (T3) and *Hernandia* (T8) (Fauvel, 1909). When Denis de Trobriand discovered the sand islet of Denis, outside the granitic group, he described it as having fine grass prairies and being half covered by big trees with trunks too fleshy and spongy for shipbuilding (Fauvel, 1909). These were probably *Hernandia* and *Pisonia* (T11), which are still present there (Fryer, 1910). Presence of *Pisonia* on Sainte Anne in 1772 is suggested by Gillot's mention of mapou among the large trees which de Launay was instructed to leave when making clearings (Fauvel, 1909).

More specific references to native trees begin shortly after colonization. Within the first five years du Barré compiled several manuscript lists of valuable native timbers; these might be suspected of being unreliable except that they agree fully with first-hand accounts written by Berthelot de la Coste, Malavois, and Rosnevet between 1781 and 1787. In addition to inland trees, these sources repeatedly mention the following as large trees growing along the shores of various islands: tacamaca — *Calophyllum* (T3), bois jaune or bois de chauve souris — *Ochrosia* (T10), two kinds of bois de rose or porcher — probably *Cordia* (T5) and *Thespesia* (T13), bois de table — *Heritiera* (T7), and pin — *Casuarina* (X1), which Malavois recognized as the same as the "pine of Madagascar" (Fauvel, 1909). The Abbé Rochon's previously noted obliviousness of everything in the Seychelles except the coco de mer is particularly regrettable here; he claimed later (Rochon, 1791) that he had introduced this *Casuarina* into Mauritius from Madagascar in 1768 and it must have been growing in the Seychelles when he visited there the following year. By 1773 it had given its name to a bay on Mahé, Anse aux Pins, and the same year du Barré had obtained an unequivocal description of it from mature trees in the Seychelles. Two other timber trees, faux gayac — *Intsia* (T9) and badamier — *Terminalia* (T12), were reported by the same sources as growing both on the coast and inland on lower mountain slopes. Malavois was amazed at the great size of the badamier, which he knew in Mauritius only as a recent introduction (Fauvel, 1909).

The early colonists thus left a reasonably good record of native coastal trees, particularly those suitable for timber. Useless shrubs and herbs were beneath their notice, and to this day many of the coastal plants are not given common names by the people who live among them. No botanist saw the islands until over half a century after settlement, when Wenceslas Bojer paid a brief visit from Mauritius. Thorough study of the flora began in the late nineteenth century. Thus a complete roster of original species requires use of indirect evidence.

The postulated list (Appendix 1) is based mainly on ecological and geographical deduction. It includes all species whose distribution patterns satisfy both of the following criteria:

1. Distribution of the species within the islands is at least partly in natural habitats. This criterion eliminates the bulk of immigrant cultigens and weeds, characteristically dependent on artificial habitats for survival. Some recent plant immigrants have become naturalized outside of such places, but the danger of accepting them as natives is reduced by the second criterion.

2. Gross distribution of the species in the Indian Ocean region is evidently at least partly ancient and natural. (Data on gross ranges are derived from many published sources, from collections in various herbaria, and from my own field observation in several parts of the Indian Ocean. Documentation is too complex to present in detail but can be supplied for particular species on request.)

A *Pandanus* or screwpine (P2) is the only endemic species on the list. It is known only from the granitic islands, except for a doubtful record on Eagle Island in the Amirantes (Gardiner, 1907; St. John, 1961). Its genus is widespread in the Indo-Pacific region and includes both far-ranging sea-dispersed

Plate 12. Boat under construction on beach overhung by coconuts and *Calophyllum*, Anse à la Mouche, Mahé

species and land-locked endemics on many islands (Stone, 1961).

All the other species have very broad natural ranges. *Smythea* (L8) is not recorded anywhere between the Seychelles and Malaysia. The other species have fairly coherent patterns. All occur from the Pacific islands through Malaysia. Except for one grass (G8), all are found on the coasts of India or Ceylon or both. The natural vegetation of Indian Ocean atolls, all the way from Cocos-Keeling and the Chagos archipelago to the outlying Seychelles, is composed almost entirely of species in this list. Not all are found on each atoll. Cocos-Keeling has only 28 of the species, most of the missing being members of the mangrove community, for which suitable habitats are lacking.

To the south and west of the Seychelles, the group begins to thin out. Each species ranges into the Madagascar–Mascarene region or the Zanzibar–East African region, but not always into both. East African mangroves are the same as in Seychelles swamps, but on African beaches, dunes, and sea cliffs the oceanic species are partly replaced by continental natives (Sauer, 1965). Thus, in supratidal species the Seychelles are more like the Pacific islands than the neighboring African coast. A hard core of nineteen species ranges, with wide gaps along the temperate and desert coasts of Africa, into the Atlantic and Caribbean regions. These pantropical species include the mangrove fern, the lippia, and about half the grasses and sedges, vines, and shrubs (F1; G3–4, 6–7; H1; L1, 3–7; S1–2, 4–5, 9–10, 13). Of the trees, only *Thespesia* (T13) is naturally pantropical, but many Indo-Pacific coastal trees and mangroves have closely related congeners in the Atlantic and Caribbean regions.

These tremendous ranges may be geologically ancient. According to Berry (1930), the Eocene Wilcox flora of the American Gulf coast included close relatives of twelve of the Seychelles coastal natives (F1; L3; M1, 5, 7; S1, 4, 9; T5–6, 11–12). Fossil relatives of the Seychelles mangroves have been found widely in Tertiary swamp deposits, including a long stratigraphic series in Borneo (Muller, 1964). Fossil coconuts are discussed below.

Such ranges, recent or ancient, obviously suggest long-range dispersal by regularly effective, intelligible mechanisms rather than by occasional accidents. The main problems are to identify the dispersal agencies and to sort out natural from human introductions.

NATURAL DISPERSAL MECHANISMS

For the bulk of the postulated aboriginal coastal species, transport by ocean currents has long been suggested and extensively discussed (Candolle, 1855; Hemsley, 1885; Guppy, 1890, 1906, 1917; Schimper, 1891; Ridley, 1930; Muir, 1937). Mangroves are characteristically viviparous, the seedlings being well developed before release from the parent tree, capable of floating indefinitely in the sea and striking root when stranded. Most of the other species have seeds that are impermeable to seawater and are either buoyant themselves or enclosed in buoyant fruits. These disseminules are commonly found in sound and viable condition among beach drift, sometimes on high-latitude coasts far from any possible source. A tangible case of effective sea dispersal was provided by the plant colonization of Krakatau, which involved 44 of these 57 species (Docters van Leeuwen, 1936) but this island is only about 25 miles from possible seed sources.

Experiments with about half the species, mainly by Guppy and Muir, have demonstrated buoyancy and viability after at least a few months' flotation in seawater, considerable variation often being encountered between individual seeds of the same species. The experiments were generally terminated without establishing maximum limits. In the case of a few species (G9; L1, 3; S6, 8–9, 11; T5, 13) experimental flotation was continued for over a year, and in one instance (S1) for over two years. It is not known how long a seed would have to float to reach the Seychelles from various possible sources; the calculation would be intricate because of variations in currents with shifts in the monsoonal wind system. However, the possibility of making such a voyage appears good for all except a few of the listed species.

The dispersal of *Casuarina* (X1) is problematical, since the seeds did not float for any significant length of time in experiments (Guppy, 1906). They may possibly be rafted within their cones or in crevices of logs and pumice blocks; after the volcanic eruption of 1883, pumice from Krakatau is known to have floated across the Indian Ocean to the Mascarenes and Madagascar within about six months (Fauvel, 1891; Muir, 1937). The blocks would have been initially sterile, of course, but they are frequently stranded and refloated. Docters van Leeuwen (1936) found *Casuarina* and *Fimbristylis* (G4) seedlings germinating on stranded pumice on Krakatau beaches.

Phyla (H1) and the whole group of cosmopolitan coastal sedges and grasses (G1–8) offer migrational puzzles, almost nothing being known about dispersal of their minute seeds. In limited experiments, I have been unable to float disseminules of any of them for any significant period. Rafting, carriage on the feet of migratory shorebirds, and flotation of living plants or rhizomes have been suggested as means of their dispersal (Schimper, 1891; Ridley, 1930; Docters van Leeuwen, 1936). The last possibility is not as incredible as it may seem because some of these plants commonly grow where they are immersed in seawater at high tide.

Pisonia (T11) is almost certainly dispersed by birds, with which it is symbiotically associated on widely scattered and often uninhabited islands. Groves of these trees are favorite nesting and roosting places for boobies and other pelagic birds and the small, viscid fruits adhere abundantly to their plumage (Guppy, 1890; St. John, 1951; Stemmerik, 1964).

In the Indo-Malaysian region, fruit bats closely related to those of the Seychelles feed on fruits of various *Pandanus*, *Calophyllum*, and *Terminalia* species, sometimes carrying them for considerable distances; the bats are known to cross up to seven hundred miles of ocean (Ridley, 1930). However, the Seychelles bats are believed to be normally restricted to the archipelago and the plant species in question (P2; T3, 12) are readily dispersed by ocean currents.

Wind dispersal of the dust-like spores of the mangrove fern (F1) almost certainly accounts for its widely disjunct, pantropical distribution. Among the seed plants, *Casuarina* (X1) and *Dodonaea* (S4) have winged fruits, but it is hard to imagine them being wind-borne for any great distance, particularly in a region without cyclones.

Capacity for dispersal by oceanic drift seems to set the aboriginal coastal flora clearly apart from the inland endemics, which characteristically lack buoyant seeds. However, it is possible that some inland species evolved from ancient coastal immigrants that lost seed buoyancy during the course of evolutionary adaptation to inland habitats, as has been suggested for the Hawaiian flora (Carlquist, 1966). Such divergent evolution might be expected if the progenitors were barely capable of reaching the Seychelles, so that any reintroductions were too rare to overcome genetic effects of local selection. Successful invasion by long-range migrants would most likely have occurred during early colonization of the islands by plants, when extensive inland habitats were open to the first-comer and not pre-empted by closed forest communities. Axelrod (1952) has made this point for the early Cretaceous intercontinental migrations of flowering plants in general.

THE QUESTION OF COCONUT DISPERSAL

A suspicion of prehistoric human introduction is inevitably raised by the early presence of coconut palms in the Seychelles. Coconut planting is certainly old and widespread in the Indian Ocean region. For example, Sulaiman, an Arab traveler of the ninth century, told of pious men who planted coconuts and dug wells on newly discovered islands in the Indian seas to provision passing ships (Chiovenda, 1921–23). Basham (1949) suggests that these men may have been Buddhists, with whom care of travelers was a religious duty. The islands mentioned may have been some of the Maldives; there is evidence that the Maldives were first settled by Singhalese Buddhists and the importance of coconuts there was reported by several medieval travelers, including Ibn Batuta (Gray, ed. and transl., 1890, Pyrard). François Pyrard, who landed there in 1602 from the wreck of a French ship, believed the Maldive coconut palms, both spontaneous and planted, were more abundant than in all the rest of the world together. Pyrard described how Maldive ships were built, fitted, equipped, and provisioned entirely of materials from these trees and then loaded with cordage, mats, sails, oil, wine, sugar, and other products, all derived from the same palm, for trade with the East and Arabia (Gray, ed. and transl., 1887–90, Pyrard). Persian and Arab ships were also traditionally built of coconut planks, sewn with coir, masted with coconut trunks, and rigged with coir cordage; these materials were commonly imported

Plate 13. Coconut plantation on extensive plateau between Grande Anse and Anse Boileau, Mahé

from the Maldives, Laccadives, or India; the coconut was known to the Arabs as the Indian nut but was grown in Arabia itself as early as the eleventh century (Hourani, 1951). The history of the coconut is much longer in India, going back to some of the most archaic post-Vedic literature (Mayuranathan, 1938).

On the other side of the Seychelles in the Arab-dominated Comoro Islands, it was reported that in 1615 entire ships complete with rigging and sails could be built and provisioned from the abundant coconuts (Terry, 1655). In the zone of Arab activity along the east African coast, Ibn Said reported certain islands near Mombasa to be celebrated for their coconuts in the thirteenth century (Freeman-Grenville, 1962); on the nearby mainland at Malindi, Vasco da Gama obtained much coir cordage in 1502 (Stanley, ed. and transl., 1869, da Gama). Da Gama's sailors were probably responsible for the name coco, applying a Portuguese word for ape or bugbear, because of the obvious resemblance of the husked nut to a hairy face (Chiovenda, 1921–23; Child, 1964; Furtado, 1964).

The Portuguese brought coconuts back to Europe before 1500 and evidently introduced the species to West Africa and

Brazil early in the sixteenth century; in these regions and in the Caribbean, coconuts were planted only rarely and locally for over a century thereafter and they were not rapidly taken into cultivation by the natives (Seemann, 1863; Chiovenda, 1921–23; Small, Safford, and Barnhart, 1929; Bruman, 1944; Lawrence, 1963; Child, 1964).

The Spanish became familiar with the coconut later than the Portuguese, primarily from groves planted by aboriginal peoples of the Philippines and Pacific islands. They evidently also found wild coconuts growing on the Pacific coast of Panama, although the much-cited account by Oviedo (Gonzalo Fernandez de Oviedo y Valdes) is a compilation from various sources, written in Spain after his return, and he confused the coconut partly with other palms (Chiovenda, 1921–23; Bruman, 1944, 1945; Hodge and Allen, 1965). Assuming that the species was prehistorically present there, it is hard to say whether it arrived with early voyagers or by oceanic drift.

Persistent reports of extensive wild coconut groves on Cocos Island, three hundred miles off the Pacific coast of Costa Rica (Wafer, 1699; Heyerdahl, 1966) are due to taxonomic confusion. A palm of the *Euterpe* group, not closely related to the coconut but easily mistaken for it at a distance, grows abundantly through the forested interior; the only true coconuts on the island are a few seashore groves that may have been planted during historic time (Stewart, 1912; Cook, 1940).

Thus, in much of its present range, including the whole Caribbean-Atlantic region and parts of the Pacific, the coconut is strictly a cultivated plant, never forming truly wild populations and only occasionally volunteering from drift nuts. Persons acquainted with the species in those areas inevitably conclude that it is incapable of natural sea dispersal. However, whether because of varietal or environmental differences, the species behaves differently in the Indian Ocean. For instance, I have seen more volunteer coconuts sprouting in beach drift on a single islet of Cocos-Keeling atoll than along hundreds of miles of Caribbean and Mexican Gulf coasts. Since it was first settled in 1825, Cocos-Keeling has been made into a vast coconut plantation by the Ross family but the original Captain Ross started with a belt of volunteer palms that fringed the atoll's beaches and that yielded a harvest of nearly one-half million nuts a year. Wild coconuts had been reported as being abundant there by eighteenth-century explorers and the atoll had already been named Cocos in 1659 (Guppy, 1890; Hill, 1929).

Other remote and uninhabited Indian Ocean islands have had similar stories. At the end of the sixteenth century, the first Dutch settlers found coconut palms on Mauritius (Sauer, 1961). The first Mauritian explorers of the Chagos archipelago, Picault in 1744 and J. M. C. du Roslan in 1771, found coconuts growing there; commercial expeditions sent out from Mauritius in the 1780's to exploit these palms were so successful that local factories for preparing copra, mills for extracting oil, and permanent settlements were soon established; the coconut gave these lesser dependencies of Mauritius the once-official and still popularly used name of the Oil Islands (Scott, 1961). Early visitors had no doubts that the groves were natural (Unienville, 1838). Early in the nineteenth century, the Chagos groves were augmented by planting but jungles of self-sown coconuts are still found there (Wiehe, 1939).

Far to the west of the Chagos, passing ships sighted coconut groves on Agalega atoll in the mid-eighteenth century; settlers arrived from Mauritius in 1809 to harvest the wild nuts and establish plantations. In 1835, when the Mauritian botanist Bojer visited the atoll, he drew a sharp distinction between the young coconut plantations and the forests of wild palms sprung from sea-borne nuts (Scott, 1961). On Providence, another western Indian Ocean atoll, a French crew shipwrecked in 1769 survived for two months on coconuts and turtles (Grant, 1801).

The presence of coconuts in all these uninhabited islands strongly suggests natural dispersal. The coconut is evidently excellently equipped for sea dispersal by the buoyant outer

Plate 14. Sparse vegetation on steep granite slopes of Ile aux Vaches Marines off Grand Anse, Mahé

Plate 15. Inner margin of mangrove swamp (Fig. 16) at high tide, L'Islette, Mahé

husk and impermeable inner seed coat, adaptations which are shared with the nipa palm of Indo-Pacific tidal swamps and many unrelated coastal plants. Experiments carried out independently in Hawaii and Celebes showed that coconuts remain buoyant and viable after several months' flotation in seawater, but no limits were established before termination of the experiments, and there were indications that seedling development actually benefitted from the treatment (Docters van Leeuwen, 1936; Edmonson, 1941; Borssum Waalkes, 1960). Edmonson concluded from his own experiments that ocean dispersal for a distance of 3,000 miles was easily possible. The recorded arrival of viable sea-drifted coconuts on the Indian Ocean island of Rodrigues (Leguat de la Fougère, 1708) would require a voyage of that order.

The native Seychelles coconuts have exceptionally thick husks and small seeds compared to common cultivated varieties (Durocher-Yvon, 1953). Unusually thick-husked varieties have also been reported from uninhabited Palmyra atoll in the Pacific and from the Nicobars and Madagascar (Beccari, 1916). Very little is actually known about the geographic distribution of wild or cultivated coconut varieties. Their classification is in a primitive state largely based on local names given to the innumerable variants that result from cross-pollination in a genetically variable population (Menon and Pandalai, 1958).

In addition to its buoyant nuts, the coconut has other adaptations suggesting long natural selection in seashore habitats: leaves that function best in stronger light and wind than most plants can endure, an extremely tough and elastic trunk, a root system that gives phenomenally firm anchorage in loose sand, ability to withstand brief flooding by seawater and to exploit effectively the fresh-water lenses characteristically present in coastal sands (Copeland, 1906). The huge food storage in the nuts may also be a natural adaptation to nourish the deep root system a seedling must develop before it can tap water to support a leafy shoot. Burkill (1935) thought it

Plate 16. Vicinity of transect (Fig. 9) on north coast of Frigate

incredible that the coconut could have evolved anywhere but on seacoasts, particularly because of its preference for loose, sandy soil aerated by rising and falling tides.

It is not crucial here whether the coconut has New World or Old World affinities (Cook, 1901, 1910; Beccari, 1913, 1917). Whatever its ultimate ancestry, the coconut must have reached the Indo-Pacific region long before the human species existed. This is the region of concentration of varietal diversity in the coconut itself and in its insect parasites; only one kind of coconut, a Panamanian variety, is peculiar to the New World (Child, 1964). Fossils of *Cocos*-like palms and of the nut itself have been found in Eocene and other early Tertiary beds in India, some described as closely resembling or identical with the modern coconut (Sahni, 1946; Kaul, 1951; Rao and Menon, 1964). A Pliocene palm fruit from New Zealand can be distinguished from the living coconut only by the relatively small size of the seed (Berry, 1926).

The most exhaustive available survey of evidence on origins of the coconut was published by Chiovenda (1921–23). After considering data from botany, ecology, paleontology, ethnology, and history, he concluded that the species evolved into its present form through natural selection on coral islands in the western Indian Ocean, among which it was dispersed by ocean currents. He believed cultivation began in the Indian Ocean region and the species was carried eastward by prehistoric peoples through the Pacific islands as a domesticated plant. This hypothesis seems to be the best available, except that it is not yet possible to clearly sort out natural and artificial dispersal in the eastward spread of the species.

If Fosberg (1962) is correct in believing that atolls were generally submerged during interglacial and postglacial high sea stands, coconut populations must have been decimated and high islands such as the Seychelles may have been vital refuges. Although best adapted to sandy beach ridges, volunteer coconuts do occur in the Seychelles on boulder slopes and crevices in the granite. As a refuge during times of near extinction, the Seychelles would have offered much less competition from other plants and less animal predation than did continental coasts. There are native Seychelles beetles which attack coconuts and which have, in fact, barred successful introduction of exotic varieties; but the native coconuts are comparatively resistant, particularly when growing on the seashore (Cooke, 1958).

POSSIBLE ANCIENT INTRODUCTION
OF OTHER PLANTS

In addition to the coconut, the postulated aboriginal flora contains numerous plants used by many Pacific and Indian Ocean peoples for boat timber, cordage, tanning, dyestuffs, food, fish poison, medicine, and ceremonial material (Safford, 1905; Burkill, 1935; Luomala, 1953; Sopher, 1965). It is easy to imagine these having been brought either by Malay voyagers en route to Madagascar or by other hypothetical prehistoric visitors (Dupont, 1938). However, in the absence of archaeologic evidence, this notion is hard to defend on purely botanical grounds. The coastal flora was not a selection of particularly valuable species but an ordinary mixture of Indo-

Pacific strand plants, useful and useless alike. It is also hard to reconcile ancient plant introductions with the initial absence of rats or feral domestic animals, which did become naturalized in the Seychelles with historic human colonization (Frappas, 1820; Fauvel, 1909). Finally, the total mystery and fabulous value of the coco de mer argue against early discovery by anyone able to return to the outside world.

POSTULATED NATURAL PATTERNS

As a working hypothesis, it is proposed here that the Seychelles at the time of colonization had vegetation as certainly virgin and natural as historic man has encountered anywhere in the tropics. In physiognomy and floristic composition, the coastal vegetation was probably very similar to that of many other Indian and Pacific Ocean islands. Quiet intertidal swamps bore a complex forest of common Indo-Malayan mangroves. Beaches and rock coasts had an outer fringe of cosmopolitan grasses and vines, backed by thickets of shrubs which were overhung by coconut palms and had tall hardwood trees at the rear. Except for the endemic screwpine on granitic glacis, all the shrubs and trees were common paleotropic species. Local contrasts in substrate and exposure produced quantitative differences in vegetational structure and composition. Details of these patterns, as shown by surviving relics, are discussed in Chapter 5.

Historic Changes in Composition of the Coastal Vegetation 4

NATIVE COMPONENT

Most of the species postulated as comprising the original coastal vegetation are still important on the coast and all but six species (enclosed in brackets in Appendix 1) were encountered in study sites of the present investigation. The last known record of *Pisonia* (T11) in the granitic islands dates from 1905 (Summerhayes, 1931) and the last of *Ximenia* (S13) from the 1870's (Baker, 1877). The other four are known to survive elsewhere in these islands, having been found as recently as 1962 by Jeffrey (collections at Botanic Gardens, Mahé). *Premna* (S7) and *Cerbera* (T4) may always have been uncommon in the Seychelles. *Ochrosia* (T10) and *Intsia* (T9), called faux gayac because its wood resembles *Guaiacum* or lignum vitae, were once abundant. Both were immediately recognized as valuable timber trees by du Barré, Malavois, and others, and were soon being selectively cut (Frappas, 1820; Unienville, 1838; Fauvel, 1909). Horne (1875) reported that *Intsia* had become very scarce and the few living trees were mostly small. He did not mention *Ochrosia* but in 1899

51

Schimper found it becoming rare (Schimper et al., 1922). Elimination of *Intsia* followed the same pattern in many other islands. It was logged out at an early date from the Chagos archipelago (Willis and Gardiner, 1931; Wiehe, 1939; Scott, 1961) and in many Pacific islands it was the most sought after timber (Burkill, 1935; Fosberg, 1960). Replanting was originally advocated by Malavois and was attempted repeatedly without success. In the present century the Seychelles Department of Agriculture turned to importing *Intsia* seed from Malaysia because pods produced by surviving native trees were destroyed by caterpillars (Dupont, 1911–22, 1923–32).

Three other native coastal trees, *Calophyllum* (T3), *Heritiera* (T7), and *Casuarina* (X1), have persisted in spite of long-continued logging.

The persistence of *Calophyllum*, or tacamaca, has been especially remarkable. Prominent in all early accounts of valuable timber trees, it was described by Bougainville as at least equal to oak in ship construction. Both Bougainville and Malavois noted its excellence for curves in naval architecture, because of the way the trees sprawl along the beach before growing vertically, and Malavois recognized the value of the straighter trunks for pirogues (Fauvel, 1909). Frappas (1820) saw a 60-foot pirogue with a 5-foot beam hollowed out of a single tacamaca. Local shipbuilding began in 1791 and was the main industry of the islands for several decades after 1800 (Toussaint, 1965). Tacamaca undoubtedly was much used in constructing some fifty vessels, of between 30 and 400 tons burden, that were launched by Seychelles shipwrights between 1810 and 1840 (Belcher, 1843; Pridham, 1846). In the 1870's it continued to be extensively used in shipbuilding for beams, planks, boards, and masts; excellent pirogues were hollowed out of the trunks, and coconut mills out of the stumps (Horne, 1875). Medium-sized vessels are being built today from tacamaca on Seychelles beaches, where the species remains conveniently available (Plate 12).

Plate 17. Fringe of natural vegetation outside coconut plantation, partly sheltered beach north of La Passe, Silhouette

Heritiera, or bois de table, was recognized immediately as a source of fine hardwood lumber and great beams (Fauvel, 1909). Horne's description of it (1875) as a large but not common tree, supplying excellent timber for house-building and furniture, is still valid today.

Casuarina, the quick-growing hardwood misnamed pin or cèdre, was recorded as a source of fuel, planking, yards, and booms by du Barré, Rosnevet, and Malavois (Fauvel, 1909; Dayer, 1962). A hundred years later, Horne (1875) found it generally used for house-building and firewood. Large trees are still widely available (Plate 10) and much used locally; there is a steady interisland trade in *Casuarina* firewood cut on La Digue and Praslin, the latter supplying most of that used on Mahé (Swabey, 1961; Lionnet, *personal communication*).

Natural coastal populations of *Calophyllum, Heritiera*, and *Casuarina* have been augmented by planting in inland clearings, a practice originally recommended by Malavois. These three, alone among the native timbers, have played a prominent role in the reforestation of crown lands (Button, 1883–87; Dupont, 1911–22; Lionnet, 1957–61; Swabey, 1961). *Casuarina* has also been planted as a support for vanilla, particularly on the calcareous plateaux.

Other native coastal trees have probably been subject to less exploitation, their timber not being especially good or durable. *Terminalia* was used for pump cylinders in Rosnevet's time (Fauvel, 1909) and for pirogues and interior boarding a century later (Horne, 1875). Huge specimens are still common and are used in a desultory way.

Utilization of the dense mangrove stand that originally fringed the northeast coast of Mahé evidently began around 1800. A 1782 map by Berthelot de la Coste shows the present site of Victoria as still being mainly mangrove swamp (Fauvel, 1909). By 1819 Frappas found the coast everywhere accessible and mangroves becoming scarce, principally through use of their ashes in soap-making (Frappas, 1820). The wood has also been used locally for buildings because of its resistance to white ants, and the bark for tanning and as a red stain for deal floors. Large quantities of mangroves have been cut in the outlying islands of Cosmoledo and Aldabra for shipment to Mahé (Dupont, 1907, 1929) but commercial cutting on the granitic islands is not recorded in recent sources. Many residual stands were destroyed in the present century by coastal road construction (S. M. Savy, *personal communication*). In less densely settled sectors, some fine stands remain that are not being intensively exploited.

Little local use is recorded of other native coastal plants. The excellent fibers of the *Hibiscus* (S5) have served for cordage (Frappas, 1820; Horne, 1875) and the species has been planted to support vanilla. The coastal screwpine (P2) may have joined its introduced congener (Pf, see Appendix 2) in supplying leaves to a local industry that produced vacoa fiber bags during the period of sugar exports (Pike, 1873). The main process by which aboriginal coastal species have lost ground has been through nonselective removal to make way for coconut plantations. Introduced exotics have played a subordinate role in occupation of coastal clearings.

PLANT INTRODUCTIONS

A total of 62 species (see Appendix 2 for list, and explanation of symbols used below) encountered in the study sites are considered exotics, either because their introduction is recorded or because it can be inferred from their confinement to cultivation and other artificially modified situations. The list is fairly complete for exotics naturalized close to the sea but not for all weeds in the interiors of coconut plantations. It gives only a sampling of the highly individualistic garden floras in the dooryards of coastal houses. Unlike most aboriginal coastal species, none of the exotics are strictly confined to seashore sites, although some are concentrated on the sandy plateaux. On the whole, they form an unremarkable assemblage of cultigens and weeds now widespread in the world tropics.

Vaughan (1937) noted that the naturalized flora of Mauritius consisted mainly of tropical American and Asian species, with comparatively few African immigrants in spite of frequent shipping contact with Africa. The same holds for the Seychelles. With a few exceptions, noted separately below, nearly all the plant immigrants probably arrived from Mauritius, where they were recorded as having been cultivated or naturalized prior to their appearance in the Seychelles (Bojer, 1837; Baker, 1877; Sauer, 1961).

A dozen of the exotic coastal species are known to have been brought in by eighteenth-century colonists. Some of these were mentioned as early crops in Chapter 2: sugarcane (Gk), banana (Pe), pineapple (Ga), tobacco (Hi), sweet potato (Lc), and cinnamon (Sb). In addition, there were several trees that Malavois noted in cultivation in 1787: screwpine (Pf), cashew (Tb), rose-apple (Tj), mango (Tk), and tamarind (Tp) (Fauvel, 1909).

Most of the other species were first recorded in the nineteenth century (Frappas, 1820; Bojer, 1837; Horne, 1875; Baker, 1877; Fauvel, 1891; Guérard, 1891; North, 1892; Brauer, 1896; Schimper et al., 1922; Summerhayes, 1931). Only two of these, star-of-Bethlehem (Hh) and bougainvillea (La), were recorded from the Seychelles before they are known to have reached Mauritius (Vaughan, 1937). Both are natives of tropical America that have become so widely distributed that their route to the Seychelles is probably not traceable. Of the whole list of nineteenth-century immigrants, 22 species were clearly introduced in cultivation, including all the vines, tall monocots, shrubs, and trees but only three herbs: taro (Gc), the

Plate 18.   Coconuts planted on almost inaccessible granite glacis, south coast of Silhouette

common garden bean (Hj), and patchouli (Hk). Like cinnamon and cashew before them, some of these escaped to grow spontaneously in disturbed habitats. *Adenanthera* (Ta) and *Pterocarpus* (Tm) were particularly quick in becoming important in secondary vegetation (Horne, 1875).

Some other nineteenth-century immigrants, 17 species of grasses and other herbs, are not known to be deliberate introductions. All are common weeds of coconut plantations and dooryard gardens, and are able to survive without planting. Some are valuable forage plants and others have minor uses in medicine or as ornamentals, so deliberate introduction cannot be entirely ruled out. Some very weedy members of this group, Bermuda grass (Gd), Guinea grass (Gh), *Asystasia* (Hd), *Stachytarpheta* (Hm), *Turnera* (Ho), and *Waltheria* (Hp), were intentionally introduced to Mauritius through botanic gardens (Sauer, 1961). In 1911, nearly forty years after it was first recorded in the Seychelles by Horne, *Asystasia* was reintroduced from India through the botanic garden at Victoria, whence it spread rapidly within a few years (Dupont, 1911–22). P. R. Dupont, then curator of the garden, regarded it as an excellent fodder plant and worth propagating under coconuts as green manure.

Dupont was responsible for many other plant introductions, visiting India, Ceylon, and Malaya for that purpose in 1901 and 1912. As director of the Seychelles Department of Agriculture from its organization in 1901 until 1932, he arranged innumerable shipments of plant material from Mauritius, India, Malaya, Hawaii, the United States, Mexico, British Guiana, Brazil, and the West Indies (Dupont, 1911–22, 1923–32, 1926). The previously noted attempts to reestablish *Intsia* with imported seed and equally abortive introductions of improved coconut varieties were made during this period. Of 15,000 Ceylon coconuts planted in 1905, a single survivor remained fifty years later, the planters since that time using native coconuts almost exclusively (Durocher-Yvon, 1953; Cooke, 1958). New varieties of Guinea grass (Gh), pineapple (Ga),

Plate 19.   Pirogue returning for another load of copra for
schooner standing off La Passe, Silhouette

Plate 20.   Wave-washed granite, fringe of spontaneous vegetation, and planted coconuts at base of barren, red hills,
Baie La Raie, Curieuse

banana (Pe), breadfruit (Tc), papaya (Tf), mango (Tk), and other economic plants introduced by Dupont may have succeeded better, although they are not distinguishable from older introductions of the same species. Species newly introduced to the coastal flora by Dupont include the Asiatic palmyra (Pb), which he brought back from India in 1901, the African *Pennisetum* or elephant grass (Gj), which he obtained via the United States in 1920, and two West Indian beach species, *Tabebuia* (To) and coco-plum (Sa), which he had in cultivation from unspecified sources in 1911. Dupont strongly recommended the last two for reforestation and erosion control, for which they have since been much planted (Swabey, 1961). They have become naturalized extensively in the interior of various islands and in a few coastal sites.

There remain seven species for which the earliest available local records are less than ten years old (Sörlin, 1957; Bailey, 1961; Jeffrey, 1963). These may have also come via Mauritius, where all but one have been present for over a hundred years (Bojer, 1837). The exception, a spider-lily (Ge) native to West Indian beaches, may have been introduced to Mauritius more recently (Sauer, 1961). It is a showy ornamental as are most of the other very recent Seychelles immigrants: the West Indian yellow-bells (Sh), the paleotropic gloriosa lily (Lb), Rangoon creeper (Ld), and *Ixora* (Sd). The other two, *Amaranthus* (Hc) and *Datura* (He), are weedy natives of tropical America.

Between the end of April and mid-June of 1963, coasts of several islands were reconnoitered extensively and twenty-five sites were studied either by transecting or sketch-mapping, or both (Fig. 2). Transects were in the form of continuous belts, 1 meter wide, at right angles to the shore and extending from the water to the margin of a homogeneous coconut plantation or other stable upland vegetation. Slopes were measured with an Abney hand level. Substrate, species present, estimated cover, and maximum plant height were recorded for each square meter along the belt. Multiple transects were run in a few topographically diverse sites. Sites were chosen to include a wide range of substrate, slope, exposure, and artificial disturbance. Sampling was not sufficiently random nor sufficiently extensive to yield accurate frequencies and the data presented are intended for rough comparisons rather than as definitive values.

No differences in coastal vegetation were noted which must be attributed to isolation of unique floras on particular

Figure 2. Location of study sites

islands. The patterns to be discussed are assumed to have been shaped by environmental selection, including human control, rather than by incomplete interisland dispersal. Even among inland natives, relatively ill adapted for dispersal, only a minority were confined to a single island, usually Mahé, the largest and most mountainous island (Summerhayes, 1931).

Supratidal and intertidal vegetation, being nearly discrete in species composition and physiognomy, are discussed separately.

TABLE 1

Most frequently encountered supratidal species, arranged in average sequence from the sea inland, according to the relative position of their seaward limits

	Outpost index	Mean presence per 10 transects
†Sporobolus (G7)	2.2	2
†Ipomoea (L7)	2.5	8
†Scaevola (S8)	2.5	8
†Pandanus (P2)	2.6	2
†Hibiscus (S5)	3.6	2
†Vigna (L9)	3.7	4
†Canavalia (L2)	4.1	3
†Cordia (T5)	4.2	4
†Guettarda (T6)	4.4	2
†Calophyllum (T3)	4.7	6
†Calonyction (L1)	4.8	2
†Cocos (P1)	5.4	9
†Barringtonia (T1)	5.5	3
†Hernandia (T8)	6.8	2
†Terminalia (T12)	7.0	2
Kyllinga (Gf)	7.0	3
Asystasia (Hd)	7.4	4
Stenotaphrum (Gm)	7.7	6
†Casuarina (X1)	8.0	2
Stachytarpheta (Hm)	9.3	4
Desmodium (Hf)	9.3	4

†Postulated aboriginal species.

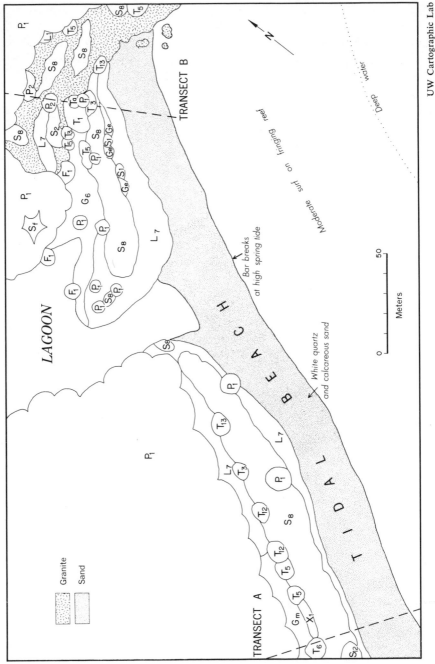

Figure 3. Species patterns, Police Bay, Mahé

UW Cartographic Lab

Species Patterns. — The supratidal vegetation of the study sites includes over a hundred species, nearly all those listed (Appendices 1–2) except the mangroves. Most of the species appear here and there on the sample-area maps and transects but they are too few and far between to be quantitatively important in the vegetation. The discussion focuses on a relatively small number of the most frequently encountered species, which are arbitrarily defined as those present in more than 20 percent of the 24 supratidal transects (Table 1). The sequence in this list depends on outpost indices, calculated by starting from the sea in each transect, numbering all species successively in the order of their seaward limits, and dividing the sum for each species by the number of transects in which it is present. The lowest possible index would thus be 1.0, indicating a species that reaches the outer limits of vegetation wherever it occurs.

This generalized sequence (Table 1) is most nearly realized on fairly open beaches with minimum human disturbance. Such a beach is found at Police Bay on Mahé (Figs. 3, 4*; Plate 11). It is exposed to moderately heavy surf and spray during onshore winds and is uninhabited, although backed by the inevitable coconut plantation. The beach morning-glory (L7) and two grass species (G3, 7) form a narrow but almost continuous outpost fringe, which pushes out during the quiet Northwest Monsoon season and is almost entirely washed away in the Trade season. These three species also colonize openings and sunny edges of the coconut plantation. *Scaevola* (S8) strongly dominates the active, outermost beach ridge, which is rarely topped by waves at extreme high tide with onshore wind. The coast moonvine (L1) and *Vigna* (L9) clamber on the *Scaevola* and on the trees behind. Several tree species, including *Calophyllum* (T3), *Cordia* (T5), and *Guet-*

*Vertical and horizontal scales are identical for all transects (Figs. 4, 6, 8, 9, 17, 20).

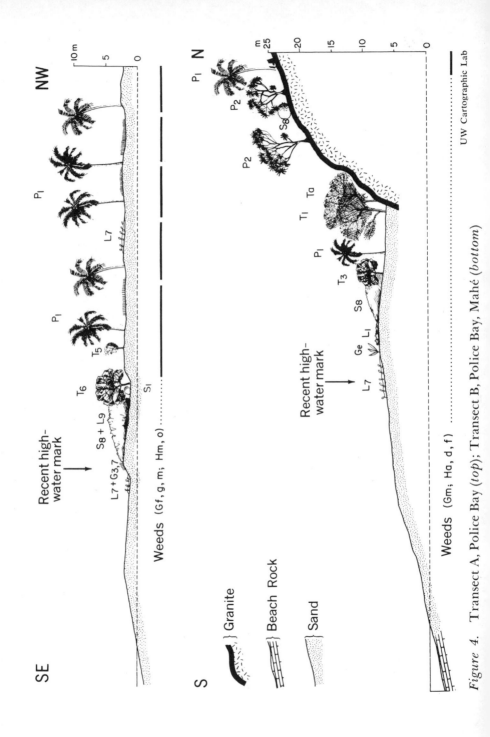

Figure 4. Transect A, Police Bay (*top*); Transect B, Police Bay, Mahé (*bottom*)

tarda (T6), form a thin row behind the *Scaevola* belt, filled in one place by planted *Casuarina* (X1). There are huge *Barringtonia* (T1) trees against the base of the cliffs at both ends of the beach. *Cocos* palms (P1) grow as scattered volunteers overhanging the outpost vegetation, well beyond the limits of the coconut plantation. Planted coconuts dominate the older beach ridges forming the calcareous plateaux and beneath them is a sparse ground cover of weedy grasses, sedges, and other low herbs. Essentially the same patterns of dominant species were found on all the open beaches studied, including parts of Grande Anse and Northwest Bay on Mahé, Grande Barbe on Silhouette, much of Grand Anse on Praslin, and the north coast of La Digue.

On sheltered beaches the species patterns are somewhat different. An example is found at Baie Ternay, in the lee of Mahé during the rough water of the Southeast Trade season and fairly well protected at all times by headlands and reefs (Figs. 5–6). The plateau is covered by the inevitable coconut plantation, through which are scattered pirogue sheds and a few estate-workers' huts with their dooryard gardens and patches of vanilla. On the outermost beach ridge and overhanging the water at high tide is a fringe of volunteer vegetation, too narrow for clear zonation of species patterns. Some of the important species are the same ones that grow behind exposed beaches: *Scaevola* (S8), *Calophyllum* (T3), *Cordia* (T5), *Casuarina* (X1), and *Cocos* (P1). They are here joined by *Hibiscus* (S5), which commonly straddles the high-tide mark, by scattered large individuals of *Hernandia* (T8) and *Terminalia* (T12) on the outer beach ridge, and by *Canavalia* (L2), which clambers on the sunny side of the whole fringe. The weedy herbs under the coconuts are mostly the same species found behind exposed coasts but here they approach closer to the shore. Very similar patterns of the dominant species are found on other sheltered, sparsely inhabited beaches, including La Passe on Silhouette (Plate 17), parts of Grande Anse on Praslin and of Baie La Raie on Curieuse, and much of the west coast of La Digue.

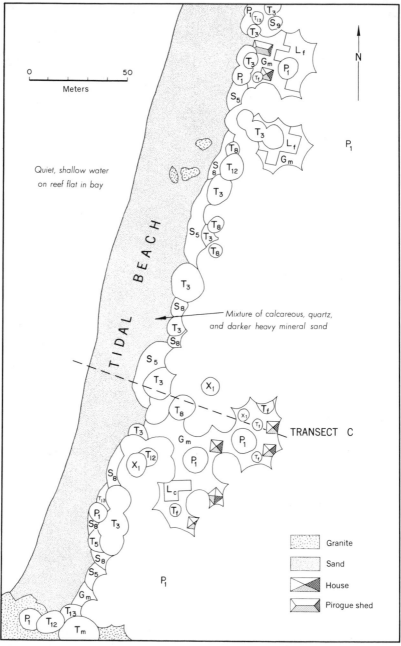

Meters

Quiet, shallow water
on reef flat in bay

Mixture of calcareous, quartz,
and darker heavy mineral sand

TRANSECT C

Granite

Sand

House

Pirogue shed

UW Cartographic Lab

Figure 5. Species patterns, Baie Ternay, Mahé

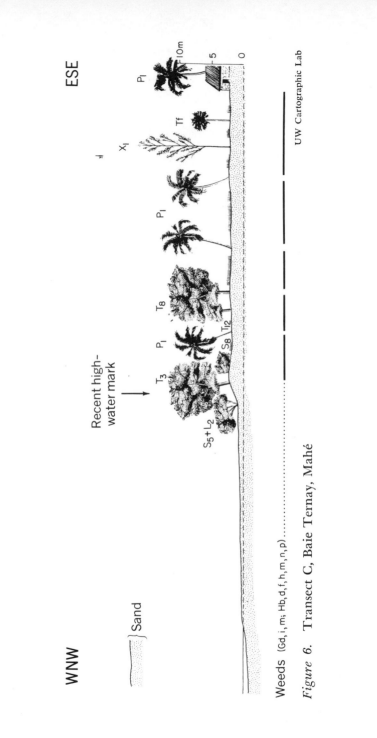

WNW

ESE

Recent high-
water mark →

⌒} Sand

Weeds (Gd, i, m; Hb, d, f, h, m, n, p) ....

Figure 6. Transect C, Baie Ternay, Mahé

UW Cartographic Lab

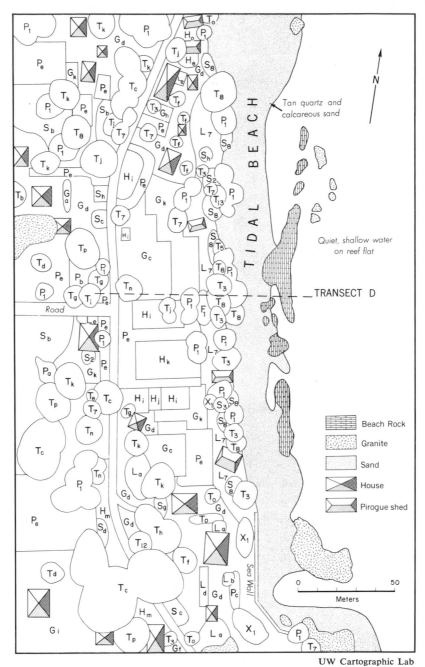

Figure 7. Species patterns, Anse aux Couches, Mahé

An example of a densely inhabited coastal sector is found at Anse aux Couches on Mahé (Figs. 7–8; Plate 8). The sector is edaphically complex, with bare granite outcropping through a series of accretion beach ridges that are composed of calcareous debris from the reef flat mixed with quartz sand. Behind the beach ridges are patches of organic soil in marshy areas, now drained by ditching. Scattered trees of *Heritiera* (T7) are evidently relics of former swampy conditions. During the Northwest Monsoon, this shore is in the lee of Mahé, and during the Trade season it is well protected by a wide reef flat and the rocky headland of Pointe au Sel immediately to the south. However, erosion is indicated by fresh exposures of beach rock. Residents report shoreline retreat of more than 10 meters during the last fifty years, and loss of a former coastal road. In spite of this attrition and intense human activity, the littoral fringe of volunteer trees and shrubs is very like that on the comparatively undisturbed Baie Ternay shore. The native *Hibiscus* (S5) is the most notable missing species. Gaps occur at pirogue sheds and along a seawall, where the natural littoral screen is replaced by a row of planted *Casuarina* (X1). Clearing behind this volunteer fringe has opened a strip for extensive colonization by the beach morning-glory (L7), inland from its natural niche in the extreme outpost zone. The inhabited area to the rear is a patchwork of small holdings, some purely residential with private gardens, others with commercial market gardens. This sector has as few coconut palms and as complex a welter of other species as any part of the coast. Similar patterns are encountered along other densely settled shores of Mahé, particularly at Anse à la Mouche and along Victoria Harbour. In detail the Anse aux Couches species mosaic is unique, although breadfruit (Tc), jackfruit (Td), mango (Tk), and tamarind (Tp) trees, garden patches of bananas (Pe), sugarcane (Gk), and taro (Gc), and weedy carpets of Bermuda grass (Gd) are common in houseyards throughout all of the inhabited islands.

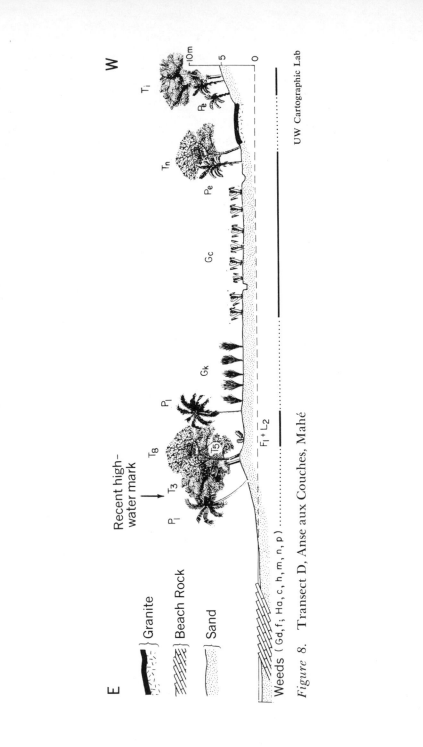

E

W

Recent high-
water mark

⟩ Granite

⟩ Beach Rock

⟩ Sand

Weeds (Gd, f ; Hɑ, c, h, m, n, p) ....

UW Cartographic Lab

Figure 8.   Transect D, Anse aux Couches, Mahé

Six transects partly or entirely on granite are included in the tabulation of commoner supratidal species (Table 1). If these and sandy sectors are tabulated separately, the sequence and relative frequencies of species remain similar, but the pattern is less consistent among individual rock transects. Accurate comparison of rock and sand areas would require more data, but most species are obviously less frequent on rock. Two of the commonest vines of sandy beaches, *Canavalia* (L2) and *Vigna* (L9), were not found in any of the rock sectors studied. *Barringtonia* (T1) seems to prefer the contact between sand and rock areas, big trees characteristically standing against granite glacis at the ends of cove beaches. The only frequent coastal species that clearly prefers granite is the screwpine (P2). Completely absent in the sandy areas studied, this species is common in rock crevices exposed to salt spray in all of the granite study sites on Mahé: the east end of Police Bay (Fig. 4 *bottom*), on Police Point, Pointe au Sel (Plates 9, 10), and on the open, granite shores of Port Launay. The species was seen during reconnaissance of Silhouette and Praslin, but does not grow in a granite study site on Praslin. It was not observed on Frigate, where exposed rocky coasts have clumps of coco-plum (Sa) and *Furcraea* (Pd) behind the usual beach morning-glory (L7) and grass zone (Fig. 9; Plate 16).

In the hope of finding unmodified rock vegetation, uninhabited Ile aux Vaches Marines was visited (Plate 14). The granite is too smooth and precipitous to give plants much foothold and the only plants in the spray zone were a solitary coconut and two grasses (G6, 8). Higher up, the total flora consisted of four noncoastal pioneer species—a sedge (*Cyperus ligularis*), a shrubby fig (*Ficus avi-avi*), a leafless milkweed vine (*Sarcostemma viminale*), and low mats of a member of the madder family (*Oldenlandia corymbosa*).

Some coastal species that are common and widespread on other Indian Ocean islands are notable by their absence or rarity in the Seychelles. The historic disappearance of a few

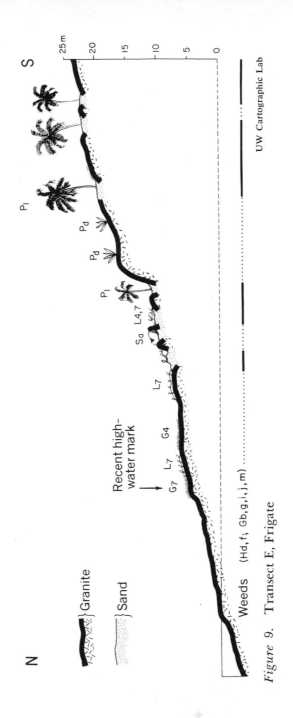

Figure 9. Transect E, Frigate

tree species has already been noted. Two shrubs, *Suriana* (S10) and *Tournefortia* (S11), which are generally associates of *Scaevola*, occur on Seychelles beaches but with surprising infrequency. *Pemphis acidula*, one of the commonest of all Indo-Pacific atoll shrubs, has never been found in the granitic Seychelles, although it is a dominant on the outlying islands. Even harder to explain is the absence of a succulent carpetweed, *Sesuvium portulacastrum*, nearly ubiquitous on tropical coasts as an extreme outpost creeper on basalt and a wide range of sand substrates.

Physiognomic Patterns. — Distributions of generalized growth-form types (Figs. 10–12) are much simpler than the underlying species patterns. The average sequence (Table 1) of outpost grasses and prostrate vines backed by a thicket of shrubs grading up to trees, with vines clambering over the shrub thicket and into the trees, is in accord with the classical zonation pattern of humid tropical beach vegetation. However, in the Seychelles the zones are unusually narrow and often tightly telescoped (Fig. 4). Instead of occupying wide storm beaches, the outpost vines and grasses expand and contract in a narrow strip kept open by changes in ordinary wave reach with seasonal wind reversal (Plate 11). The shrub and coastal tree zones, largely controlled by a gradient in salt spray, are also shallow compared to those of stormier regions.

On granite coasts, the zonation of creepers, shrubs, and trees persists as a statistical tendency, but is disrupted wherever crevices among the rocks permit the establishment of larger plants. In such places, distinctive large monocot growth-form types are added by the clumps of screwpines and *Furcraea* (Figs. 4 *bottom*, 9–10; Plate 9).

In general, however, the thickets of broadleaf dicot shrubs and trees behind beaches and of screwpines on rock form a thin screen at the outer edge of the coconut plantations (Figs. 10–11). These plantations are only exceptionally even-aged stands laid out in regular rows, but even so they look highly

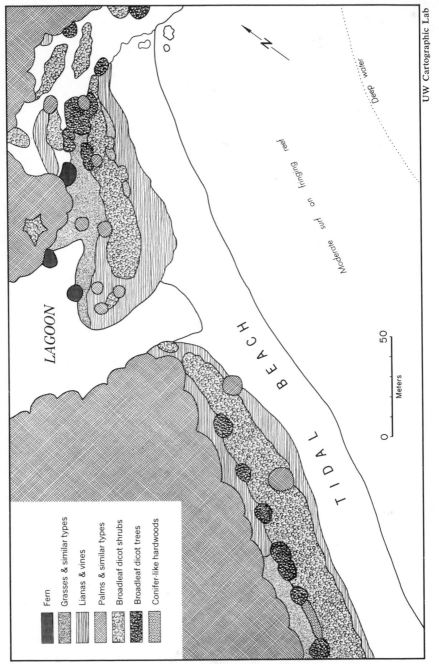

Figure 10. Physiognomic patterns, Police Bay, Mahé

LAGOON

TIDAL BEACH

Moderate surf on fringing reef

Deep water

N

0   50
Meters

UW Cartographic Lab

Fern
Grasses & similar types
Lianas & vines
Palms & similar types
Broadleaf dicot shrubs
Broadleaf dicot trees
Conifer-like hardwoods

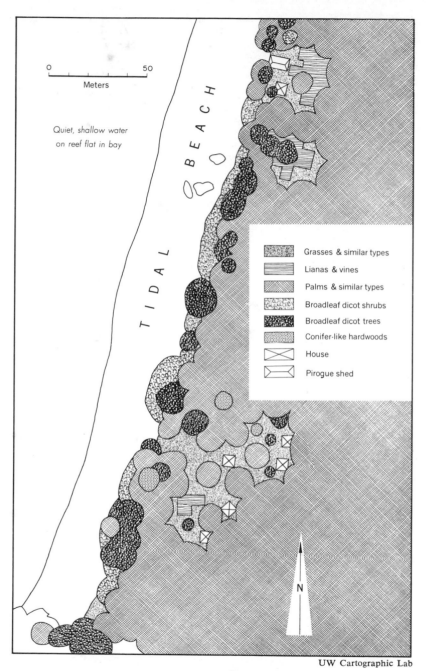

Within the figure:

Meters
0   50

Quiet, shallow water
on reef flat in bay

T I D A L   B E A C H

Grasses & similar types
Lianas & vines
Palms & similar types
Broadleaf dicot shrubs
Broadleaf dicot trees
Conifer-like hardwoods
House
Pirogue shed

N

UW Cartographic Lab

Figure 11.   Physiognomic patterns, Baie Ternay, Mahé

Legend:

- Fern
- Grasses & similar types
- Dicot herbs
- Lianas & vines
- Palms & similar types
- Broadleaf dicot shrubs
- Broadleaf dicot trees
- Conifer-like hardwoods
- Pirogue shed
- House

Quiet, shallow water on wide reef flat

TIDAL

BEACH

Road

Sea Wall

0   50
Meters

UW Cartographic Lab

Figure 12. Physiognomic patterns, Anse aux Couches, Mahé

artificial because of the monotonous uniformity of the palm canopy and the sparse undergrowth of grasses and low dicot herbs (Plate 1). Other tree types enter the picture prominently near dwellings; in densely settled sectors, vegetation becomes a physiognomic hodgepodge (Fig. 12).

Virtually the whole coastal vegetation is evergreen, exceptional deciduous members such as *Terminalia* (T12) having little effect on overall aspect. On coasts oriented so that spray intensity is variable, some seasonal upgrowth and shearing of the littoral hedge undoubtedly takes place. The bulk of the coastal vegetation is comparatively insensitive to salt spray, but inland species planted close to exposed coasts are said to become brown and defoliated during the Trade season (C. J. Piggott, *personal communication*).

Natural and Artificial Patterns. — By classifying areas on the basis of spontaneous versus planted vegetation and native versus introduced members, some other simple, general patterns are found (Figs. 13–15).

Spontaneous native vegetation persists mainly in a strip just above the normal high-tide mark. Its width is roughly proportional to exposure and inversely proportional to density of settlement, although its continuity is rarely broken even on densely inhabited sectors (Plate 8). Only the seaward borders of this type are usually purely natural, the inner ones being displaced an unknown distance by clearing and planting. Few relics of the spontaneous native forest are left inside plantation margins to delineate its former depth. However, big old *Casuarina* (X1), *Calophyllum* (T3), *Heritiera* (T7), *Hernandia* (T8), and *Terminalia* (T12) trees are still widely scattered through the coconut matrix well back on coastal plateaux, particularly on La Digue. Spontaneous native vegetation is not entirely a relic in natural habitats. Some natives are aggressive colonists of artificially disturbed sites if they are close enough to the sea. Invasions of such places by the beach morning-glory (L7) add substantially to areas

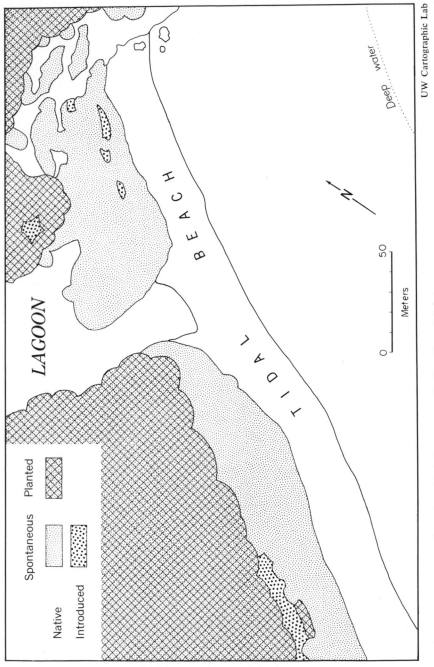

Figure 13. Natural and artificial patterns, Police Bay, Mahé

LAGOON

BEACH

TIDAL

Deep water

Spontaneous   Planted

Native

Introduced

0   50

Meters

N

UW Cartographic Lab

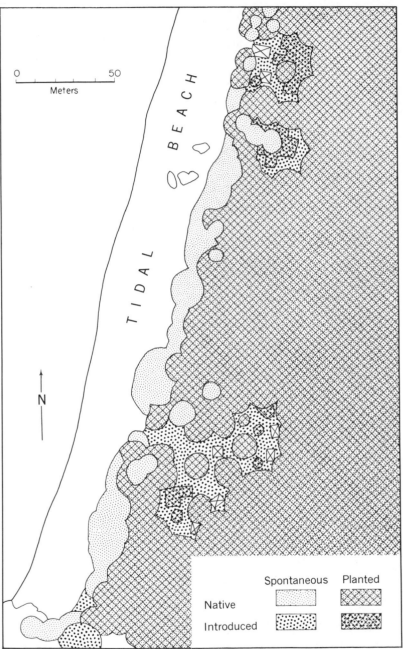

Figure 14. Natural and artificial patterns, Baie Ternay, Mahé

Within the figure:

0 — 50
Meters

BEACH

TIDAL

N

Spontaneous   Planted

Native

Introduced

UW Cartographic Lab

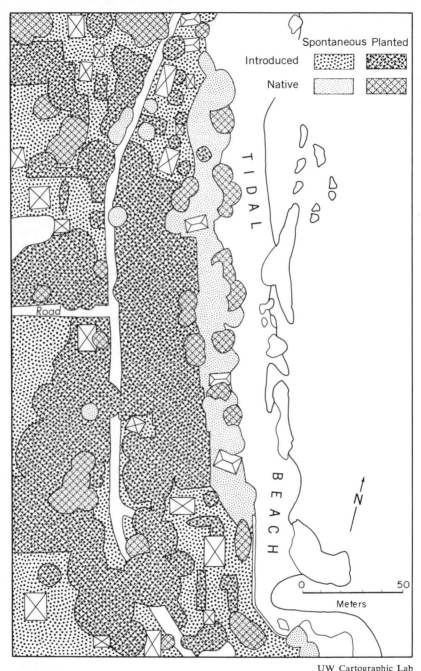

Legend:

	Spontaneous	Planted
Introduced		
Native		

TIDAL

BEACH

Road

N

0   50

Meters

Figure 15.   Natural and artificial patterns, Anse aux Couches, Mahé

mapped as spontaneous native vegetation, for example at Anse aux Couches (Fig. 15). On the site of an abandoned military installation near North Point on Mahé, an extensive clearing is carpeted by this vine (Plate 5) and by lippia (H1), another coastal native.

Vegetation composed of planted native species dominates the inner coastal zone on all but the most densely populated sectors, thanks mainly but not entirely to the coconut (Figs. 13–15; Plates 13, 18, 19). Distinguishing planted from volunteer coconuts requires local inquiry. The outpost coconuts on Anse aux Couches (Plate 8) are said by local residents to be remnants of a plantation wrecked by coastal retreat, but they look much like volunteers elsewhere (Plate 17). Likewise, *Casuarina* (X1) trees are known to have been planted near others that are spontaneous (Figs. 5, 14; Plate 10).

Spontaneous exotic vegetation occupies two quite distinct coastal situations. The rarer is in strictly natural, outpost habitats, where four tropical American natives have recently become naturalized. Examples are patches of spider-lily (Ge) on the Police Bay beach (Figs. 3, 13), a thicket of *Tabebuia* (To) near the rocky tip of Pointe au Sel, and clumps of cocoplum (Sa) and *Furcraea* (Pd) on a rocky shore of Frigate (Fig. 9; Plate 16). Spontaneous exotics occur far more extensively in weedy inner coastal situations (Figs. 13–15). The list of more frequently encountered supratidal species (Table 1) shows how regularly exotics tend to be concentrated behind the native species. They appear on the sample-area maps mainly in open spots around habitations but the same group continues as undergrowth throughout most of the area mapped as coconut plantations and extends inland indefinitely in artificially disturbed sites.

Planted exotics, none of which are strictly confined to the coast, are mostly found in dooryard gardens (Figs. 14, 15). Vanilla continues beneath the coconut canopy but it is extremely localized in distribution.

84

Wherever the intertidal zone bears vegetation, it is dominated by mangroves. Although belonging to five different families, these trees are remarkably similar in physiognomy. All have evergreen, succulent, xeromorphic leaves, grayish green and willow-like in *Avicennia* (M1), dark green and shiny in the other species. *Rhizophora* (M5) has wide-arching stilt roots, commonly forming impenetrable tangles. *Bruguiera* (M2), *Ceriops* (M3), and *Lumnitzera* (M4) have less extensive stilt roots. *Avicennia* and *Sonneratia* (M6) have shallowly buried root systems from which myriads of vertical projections arise, those of the former like asparagus stalks, those of the latter resembling miniature conical termite mounds. The more important species of *Xylocarpus* (M7) has sharply ridged horizontal roots snaking across the swamp surface. This species develops broader crowns than the other mangroves and more massive trunks, some one and one-half meters in diameter at breast height. Perhaps because of the lack of deep alluvial silt, Seychelles mangroves do not reach the huge sizes attained by the same species elsewhere; they are here usually only five or six meters tall, rarely about ten meters.

Experimental investigations with several of the same mangrove species in East Africa (Walter and Steiner, 1936) showed that their tissues have remarkably high chloride content and osmotic pressures averaging between 30 and 35 atmospheres, far above that of seawater. They are thus able to absorb water molecules by osmosis from the sea and are totally independent of any supply of fresh groundwater.

Except for a few individuals of one species of *Xylocarpus* (M8) on a sandy beach near North Point, Mahé, no mangroves were encountered above high-tide level.

Seychelles tidal swamps contain few plants other than mangroves. Epiphytes, characteristic of continental mangrove swamps, are totally lacking. Two lianas, *Derris* (L5) and

Smythea (L8), occur rarely in the depths of the swamps. The most conspicuous mangrove associate is a coarse fern, *Acrostichum* (F1), with erect fronds up to two meters tall. This fern commonly forms dense masses toward the brackish swamp margins and also in freshwater marshes subject to rare incursions of seawater (Fig. 3). Osmotic-pressure determinations for this species give values slightly below that of seawater (Walter and Steiner, 1936). A low sedge (G4) grows on open spots among the mangroves, where it is submerged by tides daily; it also occurs in the spray zone on supratidal coasts. A few primarily supratidal shrubs and trees mingle with mangroves at the inner, brackish fringes of swamps: *Hibiscus* (S5), *Barringtonia* (T2), *Heritiera* (T7), *Thespesia* (T13), and even an occasional coconut palm.

The tidal range of over a meter, unusually large for oceanic islands, creates extensive zones suitable for mangroves in two quite different topographic situations. The best-developed mangrove forests lie behind beach ridges near open stream mouths. Such swamps were studied in eight sites — Northwest Bay, Victoria Harbour, Cascade, Grande Anse, and L'Islette on Mahé; La Passe and Grande Barbe on Silhouette; and Baie Sainte Anne on Praslin. Most of these were almost pure stands of one or two mangrove species. *Lumnitzera* or *Bruguiera* are dominant in small swamps with considerable freshwater discharge, *Rhizophora* in more stagnant and saline lagoons. More complex mosaics appear where fresh- and saltwater are tidally exchanged in large volume.

An extensive swamp of this sort occupies the broad neck of an isthmus at L'Islette, connecting the mainland of Mahé to what was apparently a former islet. The isthmus has been built outward on both sides by accretion of coral sand-beach ridges and by deposits of coarse quartz sand left by the wanderings of a stream debouching from the mountains. The sample area mapped (Figs. 16–17) lies well back in this swamp, the coconut plantation in the left center and upper right being on steep granitic slopes. The coconut plantation at

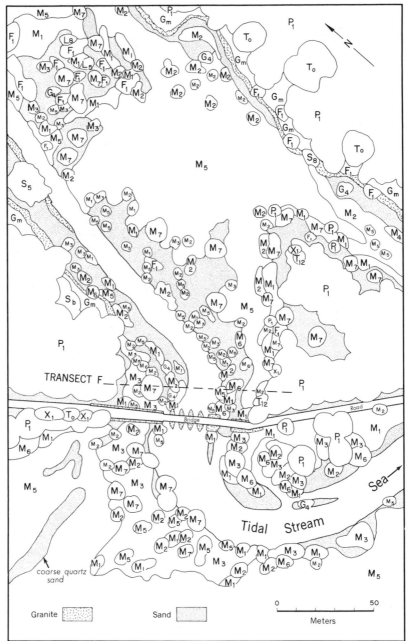

Figure 16. Species patterns, L'Islette, Mahé

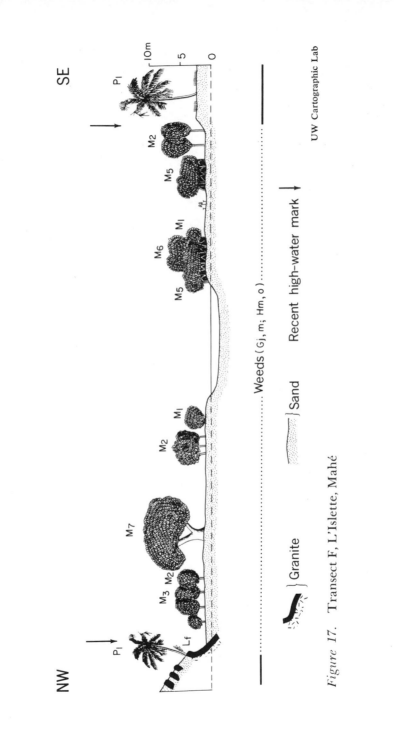

NW

SE

P₁

M₇

M₃ M₂

L_f

M₂

M₁

M₅

M₆ M₁

M₅

M₂

P₁

10m

5

0

Granite

Sand

Weeds (Gj, m; Hm, o)

Recent high-water mark

UW Cartographic Lab

Figure 17. Transect F, L'Islette, Mahé

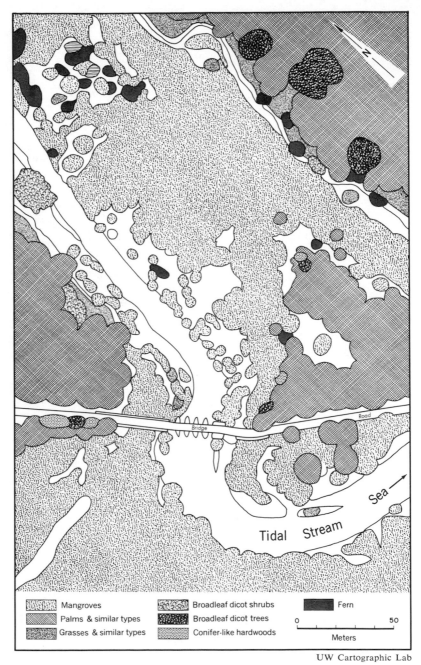

Mangroves

Palms & similar types

Grasses & similar types

Broadleaf dicot shrubs

Broadleaf dicot trees

Conifer-like hardwoods

Fern

0   50

Meters

Road

Bridge

Sea

Tidal Stream

N

UW Cartographic Lab

Figure 18.   Physiognomic patterns, L'Islette, Mahé

right center of the map is on low beach ridges. Almost all the rest of the area is flooded by tides daily (Plate 15). The relatively stagnant backswamps away from stream channels are solidly covered with a dense, pure stand of mature *Rhizophora* (M5). Clumps of small *Rhizophora* are scattered along banks of active channels. *Bruguiera* (M2), *Ceriops* (M3), *Sonneratia* (M6), and *Xylocarpus* (M7) join in apparently patternless colonization of natural levees on presently active channels and also follow abandoned stream courses in open, sandy strips that drain at low tide. *Bruguiera* is also dominant in constantly wet inner margins of the swamp where there is freshwater seepage at low tide. *Avicennia* (M1) and a low sedge (G4) colonize raw alluvium on the inside of stream bends, above and below the bridge, and also are prominent along the inner margins of the swamp. One mangrove, *Lumnitzera* (M4), a giant fern (F1), and two mangrove lianas (L5, L8), are confined to the inner margins of the swamp, as are the primarily supratidal species mentioned above as tolerating brackish conditions.

Species patterns provide the most interesting microdistributions within this swamp, the physiognomy being monotonously uniform (Fig. 18). There is no evidence of disturbance by recent timber- or bark-cutting nor are there any planted or introduced components in the tidally flooded area (Fig. 19).

The other type of topographic situation occupied by mangroves is on foreshores, open to the sea but well protected from wave attack by off-lying reefs or other barriers. Perhaps the species once grew in the regular zones noted on Tanganyika foreshores (Walter and Steiner, 1936). It was in such places that an impenetrable mangrove fringe was reported prior to settlement of the Seychelles but these quiet shores are now generally densely inhabited and the mangroves long since decimated. Along the great bay of Victoria Harbour and along Anse Boileau on the opposite side of Mahé, mangroves are still common but mostly small and widely spaced. *Rhizophora* and *Avicennia* are most abundant, with a few isolated *Sonneratia* trees standing far out on the tidal flats.

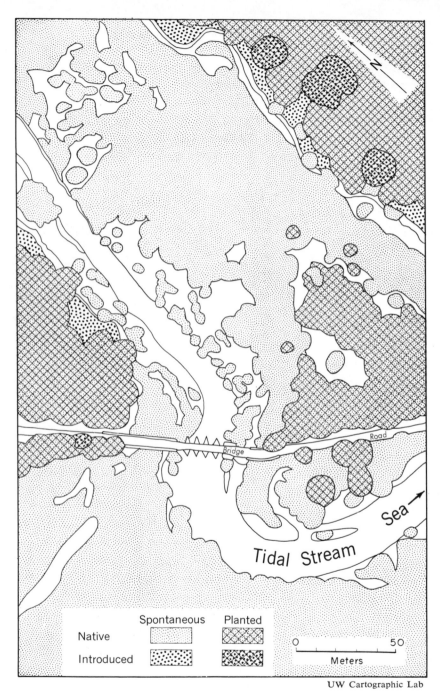

Figure 19. Natural and artificial patterns, L'Islette, Mahé

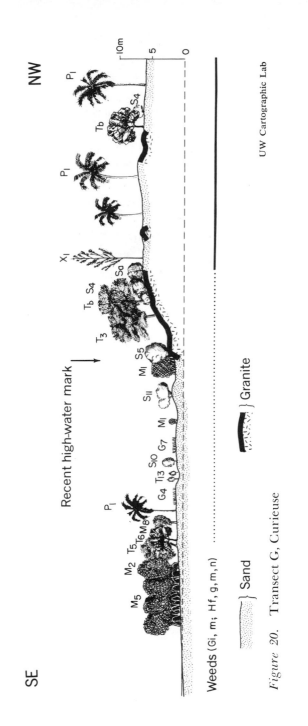

SE

NW

Recent high-water mark →

P₁   P₁   P₁

X₁

T₃   T_b   S₄   Sₐ

T_b   S₄

S₅

M₁

S_II

S₁₀   G₇   M₁

G₄   T₁₃

T₅

M₂   T₆M₈

M₅

P₁

─10m
─5
─0

UW Cartographic Lab

Weeds (Gi, m; Hf, g, m, n)

⌐‾⌐} Sand

▰▰▰} Granite

Figure 20.   Transect G, Curieuse

An exceptional case of foreshore colonization on a sparsely inhabited coast is in progress at Baie La Baie on Curieuse (Fig. 20). The shallow bay, originally open to waves from the southeast, was closed off in 1914 by a stone wall built to pen sea turtles. A dense mangrove fringe has interposed itself between the sea and relics of a beach thicket, including *Suriana* (S10) and *Tournefortia* (S11), species characteristic of very exposed coasts and highly anomalous as mangrove associates. Young trees of supratidal species are colonizing the old beach behind the mangrove fringe.

By and large, the mangroves, like supratidal coastal plants, are remarkably tolerant of human disturbance short of deliberate clearing.

Conclusion 6

Of all oceanic islands, the Seychelles may have had the longest time for vegetation to develop by purely natural immigration and evolutionary processes. When the ancient granitic mass first emerged from the sea is not known, but it was so far back in geologic time that most of it has been planed away by marine erosion. The residual islands, with rugged peaks far above any eustatic rise of sea level, must have been continuously available for plant colonization for many millions of years before the first human settlers arrived in 1770.

A unique aboriginal vegetation did develop in the mountains and interior lowlands of the larger islands: magnificent rain forests, palm groves, and screwpine thickets. These were composed mainly of endemic species, presumably derived from waif immigrants arriving by means of unlikely and seldom-repeated dispersals. Considering the mountainous, humid, equatorial environment, the inland communities were extraordinarily poor floristically. They proved highly vulnerable to human disturbance and few relics survive today.

The story of the coastal vegetation was quite different from the start. In physiognomy and floristic composition, it was originally much like that of ordinary volcanic and coral islands throughout the tropical Indian Ocean. There were extensive mangrove swamps. Open beaches and sea cliffs had complex zones of grasses, a variety of low herbs and vines, shrubs, coconut palms, and other maritime trees. Local habitat peculiarities, especially the granitic rocks, lack of storms, and high tidal range, did affect vegetational microdistribution patterns, but these effects were more quantitative than qualitative. Except for an endemic screwpine, all the coastal natives belonged to wide-ranging Indo-Pacific or even pantropical species, characteristically adapted for long-range dispersal by ocean currents. The Seychelles coasts thus had acquired a rich natural assemblage of over fifty cosmopolitan species, forming a resilient vegetation which was to be reshaped but by no means ruined by human intervention.

Entry of man into Seychelles history was exceptionally late. In spite of the antiquity of Indian Ocean seafaring, the archipelago may not have been discovered until 1609, when the crew of a wandering British East Indiaman spent a week there, gathering coconuts, catching fish and turtles, and admiring the fine timber and virgin landscape. The islands waited over a hundred years for the next explorers and remained uninhabited until the arrival from Mauritius of a few French planters and their slaves in 1770. For its first thirty years the colony was tiny and its condition precarious; it existed by shipping to Mauritius some natural products of the forests and the sea, and by trying small plantings of diverse crops. About 1800, the population began to grow slowly as cotton became a successful export crop. The cotton plantation system continued unchanged for some years after the British annexation in 1814, and by 1830 the population had increased to about 10,000 of which some 9,000 were slaves. Shortly thereafter cotton-planting became unprofitable in the face of American competition; its long-range prospects on the islands' easily eroded soils had never been good. The planters were shifting to sugarcane when slavery was abolished in 1835 and they were left without a labor force.

In the coconut, which had previously been exploited as a minor wild product, the plantation owners eventually found a new commercial crop that required relatively little labor. Extensive planting of coconuts began in the mid-nineteenth century and the palms soon blanketed the coastal sand flats and were being pushed far up the granite mountain slopes. Fruit trees, garden patches, and minor commercial crops, notably cinnamon and vanilla, have ever since been relegated to the fringes or understory of the coconut canopy. The coconut has been a godsend to the colony, permitting permanent cropping of thin and infertile soils, providing innumerable minor products for domestic use and the only major export product. Copra exports have enabled the colony to buy more food than subsistence farms could provide, and have so far supported a rapidly growing population, already exceeding 40,000, in spite of the fact that coconut acreage has long been static.

It is remarkable that this near monoculture is based entirely on native coconuts, independently domesticated in the Seychelles from wild seashore palms. Attempts to introduce improved coconut varieties from Ceylon and elsewhere have been total failures. Some native coastal timber trees have also been taken into cultivation and extensively planted in the islands. Certain other coastal species have been unintended beneficiaries of human activities, spreading as weeds in clearings and plantations. Conversely, a few coastal natives have been selectively eliminated, but these are exceptional. On the whole, native plants have been amazingly persistent, even on densely inhabited sectors of the coast.

Historic plant introductions have more than doubled the number of plant species growing on the coast. These exotics nearly all arrived as deliberate imports from Mauritius. The simplicity and narrowness of this migrational pathway has been less restrictive than might be supposed. The famous Mauritius botanic gardens were great plant clearinghouses, and from the outset Mauritian authorities supplied the Seychelles daughter colony with abundant seeds and plants, often more than the colonists were interested in growing. The colony soon had a fairly complete set of the useful and ornamental plants, mostly native to tropical America and Asia, that are commonly cultivated throughout the tropics. During the twentieth century, the Seychelles Department of Agriculture was able to add a few relatively unimportant species to the coastal flora by massive plant introductions from many countries.

The quantitative importance of introduced plants in the coastal vegetation is not proportionate to the number of species. Most are maintained only in cultivation around houses. Some have become weeds of coastal gardens and coconut plantations. Four species native to New World coasts have recently escaped from cultivation to form small naturalized populations on beaches and sea cliffs. Human dispersal of plants has played a much smaller role than might have been supposed if the vegetational history were more obscure. Had people arrived in the islands in antiquity, they would almost certainly be erroneously credited with introduction of the coconut and some other cosmopolitan, useful plants.

To a surprising extent, the present strong convergence of Seychelles vegetation with that of regions that have been settled for much longer periods of time is the product of purely local interactions between plants and man. Cosmopolitan pioneer species that were present naturally on the coast were evidently preadapted to spread in artificially disturbed habitats. They have participated to a considerable extent in the replacement of regionally peculiar forest communities by common weeds and crops.

In some ways, human-induced vegetational changes that have been spread over millennia in the world at large have been re-enacted in the Seychelles during less than two centuries and within a space of less than a hundred square miles. The islands' human population has grown with equally amazing rapidity in response to the increased carrying capacity of the new, anthropophilic vegetation. As the colonial government is well aware, the problem of potentially infinite population and fixed natural resources becomes wonderfully clear on tiny islands that are largely bare granite. In this respect, the Seychelles are already ahead of parts of the world where such problems can be postponed for a while longer.

Reference Matter

Appendix 1

English and Creole names are given only if known to be commonly used and not ambiguous. In the text, either common or generic names are followed by symbols within parentheses. Symbols consist of a capital letter indicating plant-growth form and the appropriate number in one of the nine categories below. In the following list, species appearing within brackets are recorded historically, but are not present in the study sites.

F — FERN, FOUGERE
    1. Acrostichum aureum
G — GRASSES AND OTHER MONOCOT HERBS
    1. Cyperus pennatus
    2. C. polystachyos
    3. Dactyloctenium aegyptium
    4. Fimbristylis spathacea
    5. Lepturus repens
    6. Paspalum vaginatum
    7. Sporobolus virginicus
    8. Stenotaphrum micranthum
    9. Tacca pinnatifida — Polynesian arrowroot, tavolo

H — Dicot Herb
  1. Phyla nodiflora — lippia
L — Lianas and Vines
  1. Calonyction tuba — coast moonvine, convolve
  2. Canavalia cathartica
  3. C. maritima
  4. Cassytha filiformis — liane sans fin
  5. Derris uliginosa — tuba
  6. Ipomoea gracilis
  7. I. pes-caprae — beach morning-glory, batatran
  8. Smythea lanceata
  9. Vigna marina
M — Mangroves, Mangliers
  1. Avicennia marina
  2. Bruguiera gymnorhiza
  3. Ceriops tagal
  4. Lumnitzera racemosa
  5. Rhizophora mucronata
  6. Sonneratia caseolaris
  7. Xylocarpus granatum
  8. X. moluccensis
P — Palms and Other Tall Monocots
  1. Cocos nucifera — coconut, cocotier
  2. Pandanus balfouri — screwpine, vacoa
S — Broadleaf Dicot Shrubs
  1. Caesalpinia bonduc — nicker-bean, cadoque
  2. Colubrina asiatica — bois savon
  3. Desmodium umbellatum
  4. Dodonaea viscosa — bois de reinette
  5. Hibiscus tiliaceus — mahoe, var
  6. Morinda citrifolia — Indian mulberry, bois tortue
  7. [Premna obtusifolia — bois sureau]
  8. Scaevola frutescens — veloutier manioc
  9. Sophora tomentosa — chapelet
  10. Suriana maritima — matelot
  11. Tournefortia argentea — veloutier à tabac fleurs
  12. Vitex trifolia
  13. [Ximenia americana]

T — BROADLEAF DICOT TREES
  1. Barringtonia asiatica — bonnet carré
  2. B. racemosa
  3. Calophyllum inophyllum — tacamaca
  4. [Cerbera manghas — tanghin]
  5. Cordia subcordata — porcher
  6. Guettarda speciosa — bois cassant
  7. Heritiera littoralis — bois de table
  8. Hernandia ovigera — bois blanc
  9. [Intsia bijuga — faux gayac]
  10. [Ochrosia maculata — bois jaune, bois de chauve souris]
  11. [Pisonia grandis — mapou]
  12. Terminalia catappa — Indian almond, badamier
  13. Thespesia populnea — bois de rose, porcher
X — DICOT TREE RESEMBLING PINE
  1. Casuarina equisetifolia — pin, cèdre

Appendix 2

English and Creole names are given only if known to be commonly used and not ambiguous. In the text, either common or generic names are followed by symbols within parentheses. Symbols consist of a capital letter indicating plant-growth form and the appropriate lower-case letter in one of the six categories below. In the list that follows, * = first recorded in the nineteenth century, and ** = first recorded in the eighteenth century; unkeyed species indicate those first recorded in the twentieth century.

G — GRASSES AND OTHER MONOCOT HERBS

  ** a. Ananas comosus — pineapple, ananas

   * b. Chloris barbata

   * c. Colocasia esculenta — taro, songe, l'arouille

   * d. Cynodon dactylon — Bermuda grass

     e. Hymenocallis caribaea — spider-lily, lis

   * f. Kyllinga colorata

   * g. K. polyphylla

   * h. Panicum maximum — Guinea grass, fataque

   * i. P. umbellatum — gazon trelle

    j. Pennisetum purpureum — elephant grass
** k. Saccharum officinarum — sugarcane, canne à sucre
 * m. Stenotaphrum dimidiatum

H — DICOT HERBS

 * a. Acalypha indica — herbe chatte
 * b. Achyranthes indica — herbe sergent
   c. Amaranthus dubius
 * d. Asystasia gangetica — mange tout
   e. Datura metel
 * f. Desmodium supinum
 * g. D. triflorum
 * h. Isotoma longiflora — star-of-Bethlehem, herbe poison
** i. Nicotiana tabacum — tobacco, tabac
 * j. Phaseolus vulgaris — bean, haricot
 * k. Pogostemon cablin — patchouli
 * m. Stachytarpheta jamaicensis — épi bleu
 * n. S. urticaefolia
 * o. Turnera ulmifolia — coquette
 * p. Waltheria americana — guimauve

L — LIANAS AND VINES

 * a. Bougainvillea spectabilis — bougainvillea
   b. Gloriosa superba — gloriosa lily
** c. Ipomoea batatas — sweet potato, batate
   d. Quisqualis indica — Rangoon creeper
 * e. Trichosanthes cucumerina — patole
** f. Vanilla planifolia — vanilla, vanille

P — PALMS AND OTHER TALL MONOCOTS

 * a. Bambusa sp. — bamboo, bambou
   b. Borassus flabellifer — palmyra
 * c. Cordyline terminalis — dracena, tricolor
 * d. Furcraea gigantea — Mauritius hemp, agave
** e. Musa paradisiaca — banana, bananier
** f. Pandanus utilis — screwpine, vacoa

S — BROADLEAF DICOT SHRUBS

   a. Chrysobalanus icaco — coco-plum, prune de France
** b. Cinnamomum zeylanicum — cinnamon, cannelier
 * c. Hibiscus rosa-sinensis — hibiscus, foulsapate
   d. Ixora coccinea — buisson ardent
 * e. Jatropha curcas — physic-nut, pignon d'Inde

* f.   Leucaena glauca

* g. Plumeria alba — frangipani, frangipane

  h. Tecoma stans — yellow-bells

T — BROADLEAF DICOT TREES

* a. Adenanthera pavonina — agati

** b. Anacardium occidentale — cashew, acajou

* c. Artocarpus altilis — breadfruit, arbre à pain

* d. A. heterophyllus — jackfruit, jacque

* e. Averrhoa bilimbi — bilimbi

* f. Carica papaya — papaya, papayer

* g. Citrus decumana — shaddock, pamplemousse

* h. Delonix regia — poinciana, flamboyant

* i. Diospyros discolor — mambolo

** j. Eugenia javanica — rose-apple, jamalac

** k. Mangifera indica — mango, manguier

* m. Pterocarpus indicus — rosewood, sangdragon

* n. Spondias cytherea — Otaheite apple, fruit de Cythère

  o. Tabebuia pallida — calice du pape

** p. Tamarindus indica — tamarind, tamarin

Literature Cited

(Victoria as place of publications refers to Victoria, Seychelles, in all cases.)

Anonymous. 1800. History of the Double Conspiracy. Manuscript, Government Archives, Victoria.
———. 1903. Court opinion on status of *Pas Géométriques*. Manuscript, Government Archives, Victoria.
Après de Mannevillette, J. B. N. D. d'. 1775. Le Neptune oriental. Demonville, Paris, and Malassis, Brest. 2 vol. and suppl.
Axelrod, D. I. 1952. Variables affecting the probabilities of dispersal in geologic time. Bull. Amer. Mus. Natur. Hist. 99:177–188.
Bailey, D. 1961. List of the flowering plants and ferns of Seychelles. 2nd ed. Imprimerie Saint-Fidele, Victoria. 39 p.
Bailey, L. H., and D. Vesey-Fitzgerald. 1942. Palms of the Seychelles. Gentes Herbarum 6:1–48. Bailey Hortorium, Cornell Univ., Ithaca.
Baker, B. H. 1963. Geology and mineral resources of the Seychelles archipelago. Mem. Geol. Surv. Kenya 3:1–140.

Baker, J. G. 1877. Flora of the Mauritius and the Seychelles. Reeve, London. 557 p.

Basham, A. L. 1949. Notes on seafaring in ancient India. J. Roy. India and Pakistan Soc. 23:60–71.

Beccari, O. 1913. Contributi alla conoscenza della palme. Webbia 4:143–240.

———. 1916. Note on *Palmae*, p. 44–48. *In* J. F. Rock, Palmyra Island, with a description of its flora. Coll. Hawaii Bull. 4.

———. 1917. The origin and dispersal of *Cocos nucifera*. Philippine J. Sci., Ser. C, Bot. 12:27–43.

Belcher, E. 1843. Narrative of a voyage round the world, performed in H.M.S. *Sulphur* during the years 1836–42. . . . Colburn, London. 2 vol.

Berry, E. W. 1926. *Cocos* and *Phymatocaryon* in the Pliocene of New Zealand. Amer. J. Sci., 5th ser., 12:181–184.

———. 1930. Revision of the Lower Eocene Wilcox flora of the southeastern states. U.S. Geol. Surv. Prof. Paper 156:1–144.

Boiteau, P. 1958. Contribution à l'histoire de la nation malgache. Editions Sociales, Paris. 431 p.

Bojer, W. 1837. *Hortus mauritianus*, ou énumération des plantes exotiques et indigènes qui croissent à l'Île Maurice. Mamarot, Port Louis. 456 p.

Borssum Waalkes, J. van. 1960. Botanical observations on the Krakatau Island in 1951–52. Ann. Bogoriensis. 4:5–64.

Brauer, A. 1896. Die Seychellen auf Grund eigener Anschaung. Ges. für Erdkunde zu Berlin, Verhandlungen 23:300–309.

Browne, B. C. 1963. The British contribution to the International Indian Ocean Expedition. J. Seych. Soc. 3:11–16.

Bruman, H. J. 1944. Some observations on the early history of the coconut in the New World. Acta Americana 2:220–243.

———. 1945. Early coconut culture in western Mexico. Hispanic Amer. Histor. Rev. 25:212–223.

Bulpin, R. V. 1960. Islands in a forgotten sea. Timmins, Cape Town. 434 p.

Burkill, I. H. 1935. A dictionary of the economic products of the Malay peninsula. Crown Agents for the Colonies, London. 2 vol.

Button, C. 1883–87. Colony of Mauritius, Annual report of the Conservator of Crown Lands and Forests at Seychelles. Government Printing Office, Port Louis.

Candolle, A. L. P. P. de. 1855. Géographie botanique raisonée. Masson, Paris, and Kessman, Geneva. 2 vol.

Carlquist, S. 1966. Loss of dispersibility in the Hawaiian flora. Brittonia 18: in press.

Child, R. 1964. Coconuts. Longmans, Green, London. 216 p.

Chiovenda, E. 1921-23. La culla del cocco. Webbia 5:199-294, 359-449.

Christensen, C. 1912. On the ferns of the Seychelles and the Aldabra group. Percy Sladen Trust Expedition to the Indian Ocean in 1905. Trans. Linnean Soc. London, 2nd Ser., Zool. 15:407-422.

Clarke, J. S. 1803. The progress of maritime discovery.... Cadell and Davies, London. 263 p.

Clemesha, W. W. 1943. The early Arab thalassocracy. J. Polynesian Soc. 52:110-131.

Colonial Office, Great Britain. 1949-65. Colonial reports on Seychelles for the years 1948 to 1964. H. M. Stationery Office, London, and Government Printing Office, Victoria.

Cook, O. F. 1901. The origin and dispersal of the cocoa palm. Contrib. U.S. Nat. Herb. 7:257-293.

———. 1910. History of the coconut palm in America. Contrib. U.S. Nat. Herb. 14:271-342.

———. 1940. An endemic palm on Cocos Island near Panama mistaken for the coconut palm. Science 91 (2354):140-142.

Cooke, F. C. 1958. Report on the coconut industry. Government Printing Office, Victoria. 55 p.

Copeland, E. B. 1906. On the water relations of the coconut palm. Philippine J. Sci. 1:6-57.

Dayer, L. 1962. M. de Malavois, l'histoire des concessions. J. Seych. Soc. 2:6-13.

———. 1963. Consequences de la Révolution Française et Quéau de Quinssy. J. Seych. Soc. 3:32-40.

Docters van Leeuwen, W. M. 1936. Krakatau, 1883 to 1933. Ann. du Jard. Bot. de Buitenzorg 16:1-506.

Dumont d'Urville, J. S. C. 1834. Voyage pittoresque autour du monde. Tenré, Paris. 2 vol.

Dupont, P. R. 1907. Report on a visit of investigation to St. Pierre, Astove, Cosmoledo, Assumption and the Aldabra group of the Seychelles Islands. Government Printing Office, Victoria. 51 p.

———. 1911-22. Colony of Seychelles, annual report on agriculture and crown lands. 1911–13: manuscripts in Dep. Agr. Libr., Victoria; 1914–22: Government Printing Office, Victoria.

————. 1923–32. Colony of Seychelles, annual report of the Department of Agriculture and Fishing. Government Printing Office, Victoria.

————. 1926. Reply to a questionnaire as regards the importance and usefulness of botanic gardens. Colony of Seychelles, Bull. du Dép. de l'Agr. et de la Pêche 7:1–2.

————. 1927. Le patchouli aux Seychelles. Colony of Seychelles, Bull. du Dép. de l'Agr. et de la Pêche 8:2.

————. 1929. Visit to the outlying islands by the Governor accompanied by the Director of Agriculture. Government Printing Office, Victoria. 38 p.

————. 1938. L'archipel des Seychelles, ses ressources naturelles, sa faune entomologique et son évolution économique. Gaud, Port Louis, 100 p.

Durocher-Yvon, F. 1953. The coconut industry of Seychelles. World Crops 5:437–441.

East African Meteorological Department. 1956–61. The weather of East Africa; annual report. Nairobi.

————. 1960–62. Summary of rainfall in Kenya and the Seychelles; annual report. Nairobi.

Edmonson, C. H. 1941. Viability of coconut seeds after floating in the sea. Occas. Papers Bernice P. Bishop Mus. 16:293–304.

Engler, A. 1882. Versuch einer Entwicklungsgeschichte der Pflanzenwelt, Theil II. Engelmann, Leipzig. 386 p.

Fairbridge, R. W. 1965. The Indian Ocean and the status of Gondwanaland. In M. Sears, ed. Progress in oceanography 3:83–136. Pergamon Press, Oxford and New York.

Fauvel, A. A. 1891. La flore des îles Seychelles. Compt. Rend. Congr. Sci. Int. des Catholiques tenu à Paris 1891 7:215–234.

————, comp. 1909. Unpublished documents on the history of the Seychelles Islands anterior to 1810, together with a cartography. Government Printing Office, Victoria. 417 p., 38 maps.

Fosberg, F. R. 1960. The vegetation of Micronesia. Bull. Amer. Mus. Natur. Hist. 119:1–75.

————. 1962. A theory on the origin of the coconut, p. 73–75. In Symposium on the impact of man on humid tropics vegetation, Goroka, Territory of Papua and New Guinea. Commonwealth Government Printer, Canberra.

Foster, W., ed. 1905. Journal of John Jourdain, 1608–1617, describing his experiences in Arabia, India, and the Malay archipelago. Hakluyt Soc. Works, 2nd Ser. 16:1–394.

Frappas, M. 1820. Relation d'un voyage fait à Madagascar, à Anjouan et aux Seychelles, pendant les années 1818 et 1819. Ann. Maritimes 2:229–273.

Freeman-Grenville, G. S. P. 1962. The medieval history of the coast of Tanganyika. Veröffentlichung Deut. Akad. Wiss. zu Berlin, Inst. für Orientforsch. 55:1–238.

Fryer, J. C. F. 1910. Bird and Dennis islands, Seychelles. Percy Sladen Trust Expedition to the Indian Ocean in 1905. Trans. Linnean Soc. London, 2nd Ser., Zool. 14:15–20.

Furtado, C. X. 1964. On the etymology of the word Cocos. Principes 8:107–112.

Gardiner, J. S. 1907. Description of the expedition. Percy Sladen Trust Expedition to the Indian Ocean in 1905. Trans. Linnean Soc. London, 2nd Ser., Zool. 12:1–56, 111–176.

———. 1936. Concluding remarks on the distribution of the land and marine fauna. Percy Sladen Trust Expedition to the Indian Ocean in 1905. Trans. Linnean Soc. London, 2nd Ser., Zool. 19:447–464.

Gibson, H. S. 1938. A report on the forests of the granitic islands of the Seychelles. Government Printing Office, Victoria. 50 p.

Grant, Charles, Viscount de Vaux. 1801. The history of Mauritius. ... Bulmer, London. 571 p.

Gray, A., ed. and transl., 1887–90. F. Pyrard, The voyage of François Pyrard. ... Hakluyt Soc. Works 76:1–452; 77:1–287; 80:288–572.

Grottanelli, V. L. 1955. Pescatori dell'Oceano Indiano. Cremonese, Rome. 409 p.

Guérard, P. J. 1891. Sept années aux Seychelles. Imprimerie P. H. Kober, Saint-Valéry-sur-Somme. 72 p.

Guppy, H. B. 1890. The dispersal of plants as illustrated by the flora of the Keeling or Cocos islands. J. Trans. Victoria Inst. London 24:267–306.

———. 1906. Observations of a naturalist in the Pacific between 1896 and 1899, II. Plant dispersal. Macmillan, London and New York. 627 p.

———. 1917. Plants, seeds, and currents in the West Indies and Azores. Williams and Norgate, London. 531 p.

Hakluyt, R., transl. 1862. A. Galvão, The discoveries of the world. ... Hakluyt Soc. Works 30:1–242.

Hänel, K. 1959. Madagaskar, Komoren, Reunion. *In* Deut. Afrika Ges., Die Länder Afrikas 3:1–121. Schroeder, Bonn.

Heezen, B. C., and M. Ewing. 1963. The mid-oceanic ridge. *In* M. N. Hill, ed. The sea 3:388–410. Interscience, New York and London.

Hemsley, W. B. 1885. Report on the scientific results of the voyage of H.M.S. *Challenger* during the years 1873–76. Botany 1(1):1–75; 1(4):1–333. H. M. Stationery Office, London.

Heyerdahl, T. 1966. Notes on the pre-European coconut groves on Cocos Island, p. 461–467. *In* T. Heyerdahl and E. N. Ferdon, Jr., ed. Norwegian archaeologic expedition to Easter Island and the East Pacific, 2. School Amer. Res. and Mus. of New Mexico, Monogr. No. 24. Santa Fe.

Hill, A. W. 1929. The original home and mode of dispersal of the coconut. Nature 124:133–134, 151–153.

Hodge, W. H., and P. H. Allen. 1965. Oviedo, on "Cocos." Principes 9:62–66.

Horne, J. 1875. Report on Seychelles Islands. Mauritius, Minutes of Council 10:428–445. [Reprinted in Trans. Roy. Soc. Arts and Sci. Mauritius, n.s., 9:52–83, 1876.]

Hornell, J. 1934. Indonesian influence on East African culture. J. Roy. Anthropol. Inst. Great Britain and Ireland 64:305–332.

Hourani, G. F. 1951. Arab seafaring in the Indian Ocean in ancient and early medieval times. Princeton Oriental Studies 13:1–131. Princeton Univ. Press, Princeton, N.J.

Jeffrey, C. 1962. Botanical excursions in the Seychelles. J. Seych. Soc. 2:2–5.

————. 1963. The botany of Seychelles. Dep. Tech. Coop., Kew. 22 p.

————. 1964. Coco-de-mer. New Scientist 372:34–37.

Kaul, K. N. 1951. A palm fruit from Kapurdi (Jodhpur, Rajasthan Desert): *Cocos sahnii* sp. nov. Current Sci. [Bangalore] 20:138.

Keate, R. W. 1850. Report no. 67/50 by Civil Commissioner to the Colonial Secretary, Mauritius, concerning the conditions of agriculture and labor in the Seychelles. Manuscript, Government Archives, Victoria.

Keller, C. 1898. Die ostafrikanischen Inseln. Schell and Grund, Berlin. 188 p.

Kersten, O. 1871. Baron Claus von den Decken's Reisen in Ost-Afrika in den Jahren 1862–65. Winter, Leipzig and Heidelberg. 2 vol.

Laplace, C. P. T. 1833–39. Voyage autour du monde par les mers de l'Inde et de Chine éxecuté sur la corvette de l'état *La Favorite* pendant les années 1830–32. Imprimerie Royale, Paris. 4 vol.

Lawrence, A. W. 1963. Trade castles and forts of West Africa. Jonathan Cape, London. 389 p.

Leguat de la Fougère, F. 1708. Voyage et avantures... en deux isles désertes des Indes Orientales. de Lorme, Amsterdam. 2 vol.

Lionnet, J. F. G. 1952. Rendzina soils of coastal flats of the Seychelles. J. Soil Sci. 3:172–181.

———. 1957–61. Colony of Seychelles, annual report of the Agriculture Department. Government Printing Office, Victoria. [1957 report by Lionnet and A. Jefferiss.]

———. 1958–59. Seychelles vanilla. World Crops 10:441–444; 11:15–17.

———. 1961. Seychelles cinnamon. World Crops 13:259–262.

———. 1962. Agriculture in the Seychelles, a retrospect. J. Seych. Soc. 2:14–32.

Louis, Père. 1961. Brayer du Barré. J. Seych. Soc. 1:42–53.

Loustau-Lalanne, P. 1961. Land-birds endemic to the granitic group of the Seychelles Islands. J. Seych. Soc. 1:22–31.

Luomala, K. 1953. Ethnobotany of the Gilbert Islands. Bull. Bernice P. Bishop Mus. 213:1–129.

Ly-Tio-Fane, Madeleine, ed. 1958. Mauritius and the spice trade: The odyssey of Pierre Poivre. Mauritius Arch. Publ. 4:1–148. Port Louis.

McEwen, A. C. 1961. Fragments of early Seychelles history. J. Seych. Soc. 1:7–21.

Mackay, H. 1941. The archipelago and Crown Colony of Seychelles. Can. Geogr. J. 23:300–307.

Martin, R. M. 1836. Seychelles. *In* his British colonial library: History of southern Africa 3:316–322. Mortimer, London.

Matthews, D. H. 1965. Geological history of the Seychelles. J. Seych. Soc. 4:1–7.

Mayuranathan, P. V. 1938. The original home of the coconut. J. Bombay Natur. Hist. Soc. 40:174–182.

Menon, K. P. V., and K. M. Pandalai. 1958. The coconut palm,

LITERATURE CITED

a monograph. Indian Central Coconut Committee, Ernakulam. 384 p.

Miller, J. A., and J. D. Mudie. 1961. Potassium-argon age determinations on granite from the island of Mahé in the Seychelles archipelago. Nature 192:1174–1175.

Mookerji, R. K. 1957. Indian shipping: A history of the seaborne trade and maritime activity of the Indians from the earliest times. Orient Longmans, Bombay, Calcutta, and Madras. 206 p.

Muir, J. 1937. The seed-drift of South Africa and some influences of ocean currents on the strand vegetation. Union of South Africa, Bot. Surv. Mem. 16:1–108.

Muller, J. 1964. A palynological contribution to the history of the mangrove vegetation in Borneo, p. 33–42. In L. M. Cranwell, ed. Ancient Pacific floras. Tenth Pacific Sci. Congr., Univ. of Hawaii Press, Honolulu.

North, Marianne. 1892. Recollections of a happy life. . . . Macmillan, London and New York. 2 vol.

Ommanney, F. D. 1952. The shoals of Capricorn. Harcourt, Brace, New York. 322 p.

Parkinson, J. 1640. Theatrum botanicum. Cotes, London. 1755 p.

Pelly, L. 1865. On the island of Mahi, Seychelles. J. Roy. Geogr. Soc. 35:231–237.

Pieris, W. V. D. 1961. International coordination of coconut research, p. 131–141. In J. Barrau, ed. Symposium on tropical crops improvement. Tenth Pacific Sci. Congr., South Pacific Commission, Noumea.

Pike, N. 1873. Sub-tropical rambles in the land of the 'Aphanapteryx.' Harper, New York. 509 p.

Pridham, C. 1846. An historical, political, and statistical account of Mauritius and its dependencies. In his England's colonial empire 1:1–410. Smith, Elder, London.

Rao, A. R., and V. K. Menon. 1964. Palmoxylon parthasarathyi sp. nov., a petrified palm stem from Mohgaon Kalan. Paleobotanist 12:1–6.

Ravenstein, E. G., ed. and transl. 1898. A journal of the first voyage of Vasco da Gama. Hakluyt Soc. Works 99:1–250.

Ridley, H. N. 1930. The dispersal of plants throughout the world. Reeve, Ashford, Kent. 744 p.

Rochon, A. M. de. 1791. Voyage à Madagascar et aux Indes Orientales. Prault, Paris. 322 p.

Rougon, M. L. T. 1876. Les cinquante pas du roi dans les colonies françaises. Rev. Maritime et Coloniale 51:772–802.

Rowe, J. W. F. 1959. The economy of the Seychelles and its future development. Government Printing Office, Victoria. 95 p.

Rumpf, G. E. 1741–50. Herbarium amboinense. Uytwerf, Amsterdam. 6 vol.

Russell, R. J., and W. G. McIntire. 1965. Southern hemisphere beach rock. Geogr. Rev. 55:17–45.

Safford, W. E. 1905. The useful plants of the island of Guam. Contrib. U. S. Nat. Herb. 9:1–416.

Sahni, B. 1946. A silicified *Cocos*-like palm stem, *Palmoxylon (Cocos) sundaram*, from the Deccan Intertrappean beds. J. Indian Bot. Soc., Iyengar commemorative volume, p. 361–374.

St. John, H. 1951. The distribution of *Pisonia grandis*. Webbia 8:225–228.

———. 1961. *Pandanus* of the Maldive Islands and the Seychelles Islands. Pacific Sci. 15:328–346.

Sauer, J. D. 1961. Coastal plant geography of Mauritius. Coastal Stud. Ser., Louisiana State Univ. 5:1–153.

———. 1965. Notes on seashore vegetation of Kenya. Ann. Missouri Bot. Gard. 52:438–443.

Schimper, A. F. W. 1891. Die indo-malayische Strandflora. *In* his Botanische Mittheilungen aus den Tropen 3:1–204. Fischer, Jena.

———, H. Schenck, and L. Diels. 1922. Beiträge zur Kenntnis der Vegetation und Flora der Seychelles. *In* C. Chun, ed. Deutschen Tiefsee-Expedition auf dem Dampfer *Valdivia*, 1898–99, Wiss. Ergebnisse 2(1):407–466. Fischer, Jena.

Scott, H. 1933. General conclusions regarding the insect fauna of the Seychelles and adjacent islands. Percy Sladen Trust Expedition to the Indian Ocean in 1905. Trans. Linnean Soc. London, 2nd Ser., Zool. 19:307–391.

Scott, R. 1961. Limuria — the lesser dependencies of Mauritius. Oxford Univ. Press, London. 308 p.

Seemann, B. 1863. Die Palmen. Engelmann, Leipzig. 368 p.

Small, J. K., W. E. Safford, and J. H. Barnhart. 1929. The coconut palm. J. New York Bot. Garden 30:153–161.

Sonnerat, P. 1776. Voyage à la Nouvelle Guinée. Ruault, Paris. 206 p.

Sopher, D. E. 1965. The sea nomads, a study based on the literature of the maritime boat people of Southeast Asia. Mem. Nat. Mus., Singapore 5:1–422. Government Printer, Singapore.

Sörlin, A. 1957. Om Vegetationen på Seychellerna. Svensk Bot. Tidskr. 51:135–165.

Stanley, H. E. J., ed. and transl. 1869. G. Corrêa, The three voyages of Vasco da Gama. Hakluyt Soc. Works 42:1–430.

Steenis, C. G. G. J. van. 1962. The land bridge theory in botany. Blumea 11:235–542.

Stemmerik, J. F. 1964. Nyctaginaceae. In C. G. G. J. van Steenis, ed. Flora Malesiana, Ser. I, 6:450–468.

Stewart, A. 1912. Notes on the botany of Cocos Island. Expedition of the California Academy of Sciences to the Galapagos Islands, 1905–1906. Calif. Acad. Sci. Proc., Ser. 4, 1:375–404.

Stone, B. G. 1961. Pandanus pistillaris in the Caroline Islands: An example of long-range oceanic dispersal. Pacific Sci. 15:610–613.

Summerhayes, V. S. 1931. An enumeration of the angiosperms of the Seychelles archipelago. Percy Sladen Trust Expedition to the Indian Ocean in 1905. Trans. Linnean Soc. London, 2nd Ser., Zool. 19:261–299.

Swabey, C. 1961. Forestry in the Seychelles. Government Printing Office, Victoria. 34 p.

Terry, E. 1655. A voyage to East India. Martin and Allestrye, London. 545 p.

Tonnet, A. 1906. History of the Seychelles archipelago. Manuscript, Government Archives, Victoria.

Toussaint, A. 1965. Le trafic commercial des Seychelles de 1773 à 1810. J. Seych. Soc. 4:20–61.

Unienville, M. L. A. M., Baron d'. 1838. Statistique de l'Île Maurice et ses dépendances. Barba, Fontainebleau and Paris. 3 vol.

Vaughan, R. E. 1937. An account of the naturalized flowering plants recorded from Mauritius since the publication of Baker's "Flora." J. Linnean Soc. London, Bot. 51:285–308.

Vesey-Fitzgerald, D. 1940. On the vegetation of Seychelles. J. Ecol. 18:465–483.

Villiers, A. 1952. Monsoon seas: The story of the Indian Ocean. McGraw-Hill, New York. 337 p.

Wafer, L. 1699. A new voyage and description of the isthmus of America. J. Knapton, London. 224 p.

Wallace, A. R. 1880. Island life. Macmillan, London. 526 p.

Walter, H., and M. Steiner. 1936. Die Oekologie der ost-afrikanischen Mangroven. Zeitschr. für Bot. 30:65–193.

Webb, A. W. T. 1960a. Agricultural census of the Seychelles Colony, report and tables for 1960. Government Printing Office, Victoria. 66 p.

———. 1960b. Population census of the Seychelles Colony, report and tables for 1960. Kenya Government Printer, Nairobi. 68 p.

———. 1962. Some aspects of Seychelles history. J. Seych. Soc. 2:33–43.

Wheatley, P. 1964. The land of Zanj: Exegetical notes on Chinese knowledge of East Africa prior to A.D. 1500, p. 139–187. In R. W. Steel and R. M. Prothero, ed. Geographers and the tropics; Liverpool essays. Longmans, Green, London.

Wiehe, P. O. 1939. Report on a visit to the Chagos archipelago. Manuscript, Sugar Res. Inst. Libr., Réduit, Mauritius.

Willis, J. C., and J. S. Gardiner. 1931. Flora of the Chagos archipelago. Percy Sladen Trust Expedition to the Indian Ocean in 1905. Trans. Linnean Soc. London, 2nd Ser., Zool. 19:301–306.

Index

A few crops, not members of the coastal vegetation, are indexed under their common names: cacao, cloves, coffee, cotton, lemon, maize, nutmeg, pepper, rice.